More praise for
TOUCHING A NERVE

"A trustworthy guide, [Churchland] gives comfort not by simplifying the research but by asking the right questions."
—Jascha Hoffman, *New York Times*

"Wide-ranging, insightful and provocative—a book to savor."
—*Kirkus Reviews*, starred review

"Charming, incisive, and persuasive. . . . The world is a better place for Dr. Churchland's efforts and her curiosity."
—Richard E. Cytowic, *New York Journal of Book*s

"The philosopher Patricia Churchland defies her neuro-fundamentalist reputation to present a supremely measured, sensible and readable account of the brain's role in making us who we are."
—Julian Baggani, *Observer*'s Books of the Year 2013

"Patricia Churchland . . . in this remarkably moving and deeply personal book, reminds us all that we not only have a brain and how it works, but she plumbs the depths of philosophy's biggest questions from a neuroscience perspective and thereby opens new vistas about ourselves and our humanity."
—Michael Shermer, publisher of *Skeptic* magazine, columnist for *Scientific American*, and author of *The Believing Brain*

"Fascinating. . . . Writing in a lively, down-to-earth style, [Churchland] interweaves an accessible, engrossing exposition of neuroscience with a primer on philosophical debates from Aristotle to Freud. . . . Gently but firmly brushing aside pious mumbo jumbo, Churchland embraces a scientific worldview that consoles less but illuminates more." —*Publishers Weekly*

"Patricia Churchland may be the world's leading neurophilosopher today, but she also hails from humble beginnings in rural Canada. And that plainspoken farm girl, that second self, is on full display in this beautiful, unpretentious, enchanting exploration of mind, morals, and the meaning of life."
—Owen Flanagan, author of
The Really Hard Problem: Meaning in a Material World

"*Touching a Nerve* is like a refreshing, bracing prairie breeze blowing away the cobwebs and obfuscation of so much philosophy and neuroscience. It is dazzlingly clear, down to earth, and often funny."
—Alison Gopnik,
author of *The Philosophical Baby: What Children's Minds Tell Us about Truth, Love, and the Meaning of Life*

"In *Touching a Nerve*, Patricia Churchland presents her elegant and modern understanding of the relationship between self and brain, while also explaining with a compelling personal warmth how she came to that deep understanding."
—Michael Graziano, author of *God Soul Mind Brain: A Neuroscientist's Reflections on the Spirit World*

"Accurate and commendably up to date."   —*Nature*

"Bold, deeply insightful, and biological to the core, with a warm and soothing touch of humanity."   —Joaquín Fuster,
author of *The Prefrontal Cortex*

"Will interest anyone who thinks that good philosophy needs to be grounded in good science or who is simply curious about how understanding the brain can help us make sense of the human condition. A terrific read!"   —David Livingstone Smith,
author of *Less than Human: Why We Demean, Enslave, and Exterminate Others*

PATRICIA S. CHURCHLAND

TOUCHING A NERVE

Our Brains, Our Selves

W. W. NORTON & COMPANY   NEW YORK • LONDON

Chapter 6 Epigraph reprinted from *War Is a Force That Gives Us Meaning*, by Chris Hedges. Available from Basic Books, an imprint of Perseus Books Group. Copyright © 2002.

For information about permission
to reproduce selections from this book,
write to Permissions, W. W. Norton & Company, Inc.,
500 Fifth Avenue, New York, NY 10110

For information about special discounts for bulk
purchases, please contact W. W. Norton Special Sales at
specialsales@wwnorton.com or 800-233-4830

Manufacturing by RR Donnelley, Harrisonburg
Book design by Brooke Koven
Production manager: Devon Zahn

The Library of Congress has catalogued the hardcover
edition as follows:

Churchland, Patricia Smith.
Touching a nerve : the self as brain / Patricia S.
Churchland. — First edition.
pages cm
Includes bibliographical references and index.
ISBN 978-0-393-05832-1 (hardcover)
1. Neuropsychology—Philosophy. 2. Cognitive
science—Philosophy. 3. Brain. I. Title.
QP360.C485 2013
612.8—dc23

2013007208

ISBN 978-0-393-34944-3 pbk.

W. W. Norton & Company, Inc.
500 Fifth Avenue, New York, N.Y. 10110
www.wwnorton.com

W. W. Norton & Company Ltd.
Castle House, 75/76 Wells Street, London W1T 3QT

1 2 3 4 5 6 7 8 9 0

CONTENTS

ACKNOWLEDGMENTS

PARTICULAR THANKS are owed to those friends who allowed themselves to be dragooned into reading early drafts and helping me navigate toward a final version. Especially to be thanked is the novelist and screenwriter Deborah Serra, who taught me to listen to the farm voices in my head. She prodded me to get off my academic high horse and speak plainly. Captain Dr. Dallas Boggs, U.S. Navy pilot, poet, surfer, and kayak partner, gave me insights along with many valuable corrections and a better paddling form. My siblings, Marion, Wilma, and Laird, helped me fill in details of various events, recalling on their own stories that had slipped by me. My neighbor, Dr. Sharon Boone, also doubled down on the manuscript, leaving a valuable pathologist's trail of queries and *what-abouts*. My undergraduate philosophy professor, the taciturn Don Brown, was gracious enough to read much of an early version, flawed though it was, and gave me a thumbs-up, which I read as unfettered, flamboyant enthusiasm. A great pioneer in the neurobiology of the prefrontal cortex, Joaquin Fuster, gave me pointed encouragement as well as much to think about regarding free will and self-control as he probed those same matters for his own book. Paul Churchland, as always, is thanked for jollying me along, spotting embarrassing errors, spotting even more embarrassing errors, and for giving me a beautiful bolt-hole on Bowen Island. Neuroscientists who kindly modified their figures for this volume include Leah Krubitzer, Olaf Sporns, Martijn van den Heuvel, Ken Kishida, Stanislas Dehaene, Rebecca Spencer, and Edward Pace-Schott. Thanks to Angela von der Lippe, senior editor and vice president of Norton, for her wise judgment and support.

For those of us from the farms who played in the school orchestra or joined sports teams, getting home was a problem, since the school buses had made their rounds long before we were ready to leave. This was made a little more difficult for me, because although a late bus going south was scheduled, I was often the only one heading north in my rather sparsely populated area. The solution I discovered in eighth grade was to hitchhike. The main social requirements of hitchhiking were to keep up a lively conversation with the driver. I came to realize that this meant not just babbling on, but asking questions to which the driver might have an interesting answer, allowing her or him to converse on a favorite subject. Typically, I did not know the people well, so holding up my end of the bargain—lively conversation—could occasionally get off to a somewhat lame start. The upshot was that I learned quite a lot about the history of the area, the political differences between farmers, the eccentric ways of some of its inhabitants, and what they found comical or preposterous. I also learned other things that were useful to me, such as where the old asbestos mine is located, where to find Indian paintings on the McIntyre Bluff, and where to go to see burrowing owls. Many thanks to all those who gave me a lift.

In the stories from the farm, I have changed names and identifying features, but the heart of each story remains as I remember it.

TOUCHING
A
NERVE

Chapter 1

Me, Myself, and My Brain

UNNERVED

M Y BRAIN AND I are inseparable. I am who I am because my brain is what it is. Even so, I often think about my brain in terms different from those I use when thinking about myself. I think about my brain as *that* and about myself as *me*. I think about my brain as having neurons, but I think of me as having a memory. Still, I know that my memory is all about the neurons in my brain. Lately, I think about my brain in more intimate terms—as *me*.

Every day brings new discoveries about the brain: here is your brain on drugs, here is your brain on music, on jokes, on porn; here is your brain on bragging, on hating, on meditating. Sometimes it seems that neuroscience knows more about myself than I do. On the other hand, such images are not explanations of anything. They merely correlate a psychological state—such as a feeling or thought—with a brain region of *slightly* heightened activity. These correlations do not even show that there are anatomically dedicated modules for, say, thinking about your money in the way that a car has a dedicated module that is its gas gauge. One big outstanding question is this: What are the other regions of the brain doing during these brain scans? Not *nothing*; that we know.

Certainly, these are still very early days in the science of the brain. Much, *much* more is sure to come. Is neuroscience going to tell me I do not know who I am? Should that bother me?

Some results are unnerving. Unconscious processes have been shown to play a major role in how we make decisions and solve problems. Even important decisions rely on unconscious brain activity. So you may wonder: How can I have control over a domain of brain activity I am not even aware of? Do I have control over brain activity I *am* aware of? And who is *I* here if the self is just one of the things my brain builds, with a lot of help, as it turns out, from the brain's unconscious activities?

There are other results that can perturb our peace of mind. Consider this: memories exist as modifications to the connectivity of neurons in the brain. Memories come into existence when brain cells—neurons—change how they connect to other neurons by sprouting new structure and pruning back old structure. This changes how one neuron connects to other neurons. Information about events in my life and about what makes me *me* is stored in patterns of connections between living brain cells—neurons. Memories of childhood, social skills, the knowledge of how to ride a bicycle and drive a car—all exist in the way neurons connect to each other.

Here is the implication that may not be welcome: In dementing diseases and in normal aging, neurons die, brain structures degenerate. In death, brain cells quickly degenerate, with massive loss of information. Without the living neurons that embody information, memories perish, personalities change, skills vanish, motives dissipate. Is there anything left of me to exist in an afterlife? What would such a thing be? Something without memories or personality, without motives and feelings? That is no kind of *me*. And maybe that is really okay after all.

Getting accustomed to the science of the brain can be a vexing business. For example, a renowned if rather melodramatic philosopher hoisted himself up in a conference I attended and, hands gripping the chair in front of him, hollered to the hushed crowd, "I hate the brain; I hate the brain!" Whatever could he mean?

He may have meant many different things, and perhaps he himself was not altogether sure exactly what he meant. Maybe he just needed to vent. Perhaps he simply felt exasperated by the blossoming of a field where his philosophical ideas had begun to seem outdated. On a more cynical take, perhaps he just wanted to draw attention to himself.

Yet he could have meant that neuroscience is mining results that he does not know how to absorb into his usual way of thinking. I am reasonably certain that he does not believe in a soul apart from the brain—*that* is not the problem. Did he perhaps mean that he does not want to learn about the mechanisms underlying his thoughts and attitudes—mechanisms that involve cells and chemicals interlocked in causality? Each of us thinks we know ourselves better than anyone else can know us. But if the *un*conscious brain is a major factor in what we think *at this very moment* and what we feel *at this very moment*, the ground under our feet may seem to be falling away.[1]

Maybe my friend found himself gravitating to the anti-enlightenment view that the mystery of the brain and how it works is better left alone. Maybe he feels that when it comes to the brain, *not* knowing is more valuable than knowing. Does he worry that neuro knowledge is forbidden fruit, a Promethean fire, a Pandora's box, a Faustian bargain, an evil genie released from a rightly sealed bottle? What sense would that codswallop make?

Deep resistance to knowledge that betokens a change in a whole way of thinking has a long history. Think only of the horror displayed by the cardinals in Rome when Galileo discovered the moons of Jupiter with his amazing new tool, the telescope. The cardinals refused to even take a look. Galileo realized that Jupiter's moons are circling Jupiter and that Venus is circling the sun, and so . . . crikey . . . probably Copernicus was right. Earth is also circling the sun—meaning that Earth is not the center of the universe.

Not long after his announcement, Galileo was placed under house arrest and forced to recant his hypothesis that Earth revolves around the sun. His decision to recant had been influenced by his having been "shown the implements of torture." So what was the big deal about Earth circling the sun? Every schoolchild learns that now and it does not stir up a hornet's nest.[2]

Why did the cardinals care so much? Did they "hate" Earth's revolution about the sun? The general answer is that they cared because of what they believed about the physics of the universe. The conventional wisdom of the day assumed that Earth is at the center of the universe. Everything below the level of the moon is corruptible, changeable, earthly, imperfect. That is the realm of *sublunar physics*. Everything above the level of the moon is perfect, heavenly, unchanging, and so forth. This is the realm of *supralunar physics*. Different laws were thought to apply. The stars were widely thought to be holes in a huge sphere (literally made of crystal) that enclosed the universe—Earth at the center, of course. This cosmology was derived from biblical text.

Copernicus and Galileo threw that cosmology under the bus. The moons of Jupiter looked pretty much like our moon, meaning that they might be dirt balls, too. And so Jupiter might just be like Earth. But then, did God *not* create Earth at the center of the universe? Does this mean that the crystal sphere does not exist? But where, then, is heaven, if not just above the moon? Where did Jesus go when he ascended bodily into the sky? A specific, long-held worldview was fundamentally challenged, and the challenge generated fear of what might replace that worldview. The very institution of the Christian Church was founded on the belief in the bodily resurrection of Jesus into an actual place—heaven. And this actual place is above the moon, and maybe even above the stars. If you believe something to be absolutely certain and foundational, it is profoundly shocking to find that your "truth" may be mushy, or worse, quite likely false.

Think next of the impact of the discovery by the English scientist William Harvey in 1628 that the heart is actually a pump made of muscle. Consternation! Say it is not so! Not a mere meat pump! What was the big deal about the heart?

The conventional wisdom of Harvey's time accepted a very different story that had been proposed by the Roman physician and philosopher Galen (129–199 CE). Galen's idea was that being alive entails animal spirits vivifying the body. Where do the animal spirits come from? They are continuously made in the heart. That is the job of the heart—to concoct animal spirits. *Vivifying*? Means *keeping alive*. So it was a rather circular and unhelpful explanation after all.[3] In any case, the idea was that the animal spirits are continuously put into the blood by the heart, and the heart constantly makes new blood.

Harvey's discovery that the heart is really a pump acknowledged that while a living animal is indeed different from a dead animal, spirits are probably not what make the difference. Blood is made somewhere else, and the heart merely circulates the blood.[4]

Harvey's colleagues were of course deeply steeped in the unquestioned "truth" of Galen's account of animal spirits. Upon seeing Harvey's data, they did in effect cry painfully, "I hate the heart, I hate the heart!" What was actually said was in a way worse. They said they would "rather err with Galen than proclaim the truth with Harvey."[5] This is the familiar strategy of *let's pretend*. Let's believe what we prefer to believe. But like the rejection of the discovery that Earth revolves around the sun, the *let's pretend* strategy regarding the heart could not endure very long.

Why the anxiety about Harvey's discovery that the heart is a pump? Because it was not merely the discovery of a little fact about an organ in your chest. For those alive in the seventeenth century, it challenged a whole framework of thinking about spirits and life that had been taken for granted as true since about 150 CE. It threatened the tight connection between the religious framework of life as a matter of spirits and the scientific framework that explored the nature of those very spirits. After Harvey, after Copernicus and Galileo, that connection ceased to be conveniently tight. Religion could either drop dogma and go with science, or religion and science would move apart.

What Harvey's discovery did do was open a whole range of new questions: Why does the blood circulate? Where does it get made? What *is* blood really? Why is some blood more brilliantly red than other blood? What *is* the difference between being alive and being dead? Suddenly what everyone supposed they knew was replaced by questions rather than answers, which ultimately led to much deeper answers, and yes, even more questions. Uncertainty, a tonic to some, is toxic to others. Trumped-up certainty, a balm to the gullible, is anathema to the skeptical.

Harvey's discovery does not seem threatening to us because we are totally accustomed to the idea that the heart is a pump that circulates the blood. Additionally, no one talks about animal spirits in Galen's sense anymore. Being alive has a completely different and very biological basis that does not invoke the infusion of animal spirits. The example is instructive nonetheless because what does seem like a big deal to *us* is the realization that we are what we are because our brains are what they are. This, too, challenges a worldview and provokes some to retreat back into the verities they wish were true. They hate the brain.

Advances in understanding the brain certainly entail thinking about ourselves in a somewhat new way. For example, it can come as a surprise to appreciate how very biological we are and how very biological our psychological processes are—how they can be affected by hormones and neurochemicals. Just as all mammals have hearts that are very similar to human hearts, all mammals have brains with much the same organization and anatomy as human brains. Several hundred years hence, students reading the history of this period may be as dumbfounded regarding our resistance to brain science as we are now regarding the seventeenth-century resistance to the discovery that the heart is a meat pump.[6]

Whatever the new scientific discovery that alarms us, a strategy of denial of the discovery cannot endure very long. The Hungarian physician Ignaz Semmelweis (1818–1865) showed that washing the hands with chlorinated lime to disinfect after dissecting cadavers and before examining women in the obstetrics ward could reduce mortality of women dying from infection from as high as 35 percent down to 1 percent. Many of his colleagues were completely offended at the very idea that they should wash their hands. Semmelweis was ridiculed as a crank and treated very badly within the medical community, despite his efforts to persuade physicians to simply test the efficacy of hand washing. Many years after his death at 47, disinfecting the hands became the norm, partly as a result of corroborating data from Pasteur and Lister. Also important was their discovery of infection-causing microbes not visible with the naked eye but visible under the microscope. Microbes explained why hand washing could prevent the spread of infection. Invisible microbes were washed off the hands.

Some 400 years after Galileo's arrest, the Catholic Church conceded what every moderately well-educated person knew to be true: Earth is not the center of the universe. It is part of the solar system which itself is way out on an arm of a minor galaxy. Contrast the *let's pretend* approach with that of T. H. Huxley: "My business is to teach my aspirations to conform themselves to fact, not to try and make facts harmonize with my aspirations."[7]

If you prefer some mystery to encircle the stuff of self, there may be comfort in the fact that vast numbers of questions about the brain, including fundamental questions, remain unanswered. For none of the higher functions—including retrieval of autobiographical memories, problem solving, decision making, consciousness, why we sleep and dream—do we have anything close to a complete and satisfying neural explanation. We have no idea at all of the neurobiological differences whereby one person is thrifty and another person a spendthrift; why one person finds math easy and another struggles; why one person is vindictive and another easily forgives. Neuroscience has many fragments and pieces of explanations, but the complexity of the brain's functions is truly daunting. Some people say that the human brain is the most complicated thing in the universe. Perhaps that is true, though how would we know? The universe is pretty vast and has hundreds of millions of galaxies. For all I or anyone else knows, in some other galaxy there are things that are vastly more complex than even human brains.

To go back to my friend's heartfelt philosophical howl, one thing did give me pause. Even a highly educated and broadly read person can feel resistance when new knowledge takes us into factual spaces that we did not know were even there—factual spaces that are hard to assimilate into the comfortable world we feel we know. We saw this with the reactions to Galileo and Harvey. This was also true regarding biological evolution, which certainly shocked many people and in some quarters still does. Acceptance of Darwinian evolution is by no means as far along as acceptance of Harvey's discovery. Accepting emotionally what you apprehend cognitively is often a tough sled. Our brains are wired that way.

Here is something that can bother people and definitely bothers me. Some hypotheses about the brain are marketed by self-promotional writers who exaggerate what we actually know, thus causing a sensation. They claim certain results to be momentous and established, as weighing high on the *gazowie!* scale. We are regaled: "free choice is an illusion," "the self is an illusion," "love is just a chemical reaction." Just how well supported are these claims? How much is marketing and manipulation?

In my judgment, such startling claims are more sensational than they are good science. They may contain a kernel of genuine evidence, but they stretch out of shape what is actually established—so much so that the kernel of truth is swamped by hype. Neurojunk misleads the same way that the claim that biological evolution entails that your grandfather was a monkey misleads. That claim distorts what we know to be true about biological evolution and slaps on rubbish. When you look closely, some of these over-egged ideas about the brain turn out to rest on modest, ambiguous, and hard-to-interpret data, contrary to what the dolled-up advertising copy suggests.

Still in "I hate the brain" territory, an entirely different worry is expressed by a few philosophers. They are exasperated by developments in neuroscience and psychology because these fields encroach on what the philosophers think of as their turf. Many contemporary philosophers, both in America and in Europe, trained for their jobs expecting to address questions about the nature of consciousness and knowledge and decisions without having to learn any neuroscience. Or *any* science, for that matter. They want to garner insights from the great books or from their own reflections. That is the true "philosophical method," they complain. "Why bother with the brain? Can we not just plumb the deep questions without having to think about the brain?"

This response is largely based on fear of job loss. You can readily sympathize with that response without wanting to turn back the clock. When cities became wired for electricity, the lamplighters had to find work elsewhere; as horses were replaced by cars, the village smithies had to learn to repair internal combustion engines. There are many things for philosophers to do, including collaborating with scientists and keeping alive the wisdom in the great books.[8] But if they want to address how the mind works, they need to know about the brain.

Putting aside neurojunk and putting aside territorial worries, there remains the substantive neuroscience that bumps right up against who we think we are and implies that we might be something a little different. That is the domain I want to talk about.

Having taught neurophilosophy to undergraduates at the University of California San Diego for many years, I understand well that the sciences of the mind/brain can be unsettling. Neurophilosophy, as I have described it, works the interface between philosophy's grand old questions about choice and learning and morality and the gathering wisdom about the nature of nervous systems. It is about the impact of neuroscience and psychology and evolutionary biology on how we think about ourselves. It is about *expanding* and *modifying* our self-conception through knowledge of the brain.

Very often, my undergraduate students were drawn to the sciences of the brain as to a night bonfire in the woods. At the same time, they also expressed apprehension concerning what changes in understanding they might bring. There is tension. There is ambivalence. What if we do not like what we come to understand? Will new knowledge make us miserable? What if we get burnt by that bonfire? Change per se can be unsettling. And restlessness provoked by neuroscience is what Owen Flanagan and David Barack call *neuroexistentialism*.[9]

Like anyone else, I find that there is much in life that can make me lose my balance, much that can provoke disharmony and distress. On the whole, however, I am not rattled by findings in neuroscience. Surprised, often; but alarmed? Not so long as the science is solid science.

Biology reassures me. The connection, via evolution, with all living things gives me a reassuring sense of belonging. I suppose it need not, but it does. It gives me a sense of my place in the universe. I watch a fruit fly and think: the genes that built it a front and a rear are essentially the same as the genes that built me a front and a rear. I watch the blue jays fly off with peanuts to cache in the woods. I think: the brain that sustains their spatial memory for where they hide nuts works pretty much the same way as my spatial memory does. I see a sow tend her piglets and I think: much the same system has me tending my children. I find joy in commonplace mental events, such as a many-factored decision that I have mulled over for days coming to consciousness one morning as I stand in a hot shower. My brain has settled into a choice, and I know what to do. Yay brain!

So how did I get here? What was the path whereby I ended up feeling comfortable with my brain, or, more exactly, comfortable with what neuroscience teaches me about my brain? There are two parts to the story. One is brief and contains the basic background logic. I doubt this is the sum and substance of the answer. But it is how, from the perspective of hindsight, I give a meaningful organization to the more indirect and vastly messier part of the answer. The messy part has to do with slow cognition, temperament, growing up, learning, role models, life experiences, successes, failures, and luck. The background logic is not unimportant here, but it is in mixed company.

To relate how I became accustomed to thinking neurobiologically, I have to tell you a somewhat personal story, not a textbook-style story. There are brilliantly written and scientifically outstanding textbooks on nervous systems, including introductory ones.[10] I am not aiming to duplicate those achievements. I am aiming to do something softer so that you can see what the world of brains looks like from my perspective. In talking about the brain, I sometimes find it useful to interweave the science with the stories. Because life on the farm was a major force shaping how I go about things, many of the stories are from the farm.

On the farm where I grew up, learning how things work was

no mere frill. Things often had to be fixed or even fashioned, and knowing how to fix or fashion a device yielded the relief of saving time and toil. Knowing how to get a kerosene lamp going during a storm or how to set a proper gopher trap was highly useful for lighting the dark and controlling the vermin. Failing with one strategy meant searching for a better way. Finding a solution could also be a source of pride. I often trotted around after my father, watching with fascination as the local tinsmith turned out new stovepipes or the juice factory turned out bottles of apple juice. When I began to learn about the brain systematically, the project seemed like more of the same—finding out how something works.

The background logic has three main points and essentially goes like this: First, reality does not conform to what we want it to be. The facts are the facts. Reality does not care if we do not like the way it is put together. It goes right along being reality anyhow. Reality does not care if we prefer to disbelieve facts about the heart or the brain or the cause of AIDS. By working with reality, we can sometimes change it by finding a new vaccine or a new machine to harness electricity. Science—testing, being guided by the facts, revising, testing again—is the best deal we have for getting a bead on reality. And this is as true on the farm or in the forest as in the lab.

Second, liking what is true is a psychological state. You can fight reality, hoping your fantasy will prevail, or you can decide to make your peace with reality and come to like it. Genuinely like it. It took the Catholic Church 400 years, but it, too, made its peace with Galileo's discovery of the moons of Jupiter and seems now to like that reality quite well. In general, fighting the truth about reality does not work out well in the long run. Like freckles: If you have them and you go through life hating them, you simply make yourself miserable. If you make your peace with your freckles, if you come to see what is charming about them, you manage much better.

And third, we can regulate how we use science. We need not heed the romantics who insist that the old days were really the good old days. The old days also had plague and no anesthetics. Nor need we heed the alarmists who try to scare us by saying that Armageddon is at hand because we know too much. Bosh. Both are opposed to enlightenment, and both are misconceived. Once at a bioethics meeting in D.C., an anti-enlightenment member wagged his finger at me and said, "You are an optimist!" His tone and the wagged finger made it evident that this was intended as a rebuke, but to me an odd sort of rebuke—first because I thought of myself as a hard-assed realist rather than a cock-eyed optimist, and second because I had not for a moment considered realistic optimism to be a vice.[11] Surely not a vice like, say, neglecting pets or gambling to excess or molesting children. Or drubbing enlightenment.

He was right about this much: Any discovery can be used for nefarious purposes, whether it is in chemistry or archaeology or metallurgy. For example, fire can be used to burn down other people's houses. And religion, too, can be used for nefarious purposes. Nevertheless, we can regulate how we do things. In the case of scientific discoveries, we can regulate the uses to which the discovery is put. To the degree that I am optimistic, it is because there are scientific discoveries that obviously and unequivocally have been used to make life better—such as polio and smallpox vaccines; such as Prozac and lithium; such as hand washing by surgeons and the use of local anesthetics by dentists. Did the finger-wagging bioethicist not know all about those developments? Yes, indeed he did, but he was ambivalent about *progress*. My sense is that he looked down on creature comforts as merely *this-worldly*. If, like me, however, you were shaped by toil on a poor farm, you are disinclined to turn up your nose at progress and creature comforts and time to read. How fine a thing is an indoor toilet on a winter morning when it is -10° Fahrenheit and there are 2 feet of fresh snow.

A quick look at the logic describing a discovery is not enough to sign on to a momentous and new way of understanding yourself and the world. It takes time to see how the logic plays out and whether there are counterarguments that weigh against the claims on offer. It takes time to see whether the new framework is coherent and stands up to heavy weather. Caution and questioning are always appropriate. The logic is like a seed, only a seed. We want to see whether the hypothesis takes root and grows into something more connected, trustworthy, and sturdy or whether the hypothesis withers, failing to explain or predict anything much and failing to compete successfully with more fruitful hypotheses.

Abrupt changes of deep beliefs can be too hasty. I had an aunt who became a communist, then dropped that after a few years and became a Mormon. In between there was a short conversion to Christian Science, brought to a halt when an abscessed wisdom tooth failed to respond to prayer and needed the attention of a dentist. Her Mormonism lasted for a brief spell, but then she scuttled that when she found the faith of the Jehovah's Witnesses. Each time her enthusiasm for her new faith knew no bounds, her certainty and conviction was unshakable, until it was. It would have been amusing except that her impulse to convert all of us made her a pest. This sort of instability about life-guiding ideologies is unusual and unwise.

There are some things that brains do very slowly and that involve deep intelligence or deep shifts in worldview. Human brains do figure things out, sometimes only after years and years of pondering and marinating and seeing what makes sense. The tale that the idea of gravity occurred to Newton one sunny fall afternoon when an apple fell on his head is bogus. He had been chewing and chewing and *chewing* for decades on the nature of motion, both of the planets and here on Earth. There is much to ponder in the developments in neuroscience, and a brash little argument that concludes that humans have no control over their lives should be treated with some skepticism before we let it worm its way into our belief system.

The slow dawning of deep ideas by human brains stands in stunning contrast with electronic computers. Computers can do many things much faster than we can, such as calculate. But computers—so far, anyhow—cannot do these deep things that human brains do slowly. They do not come up with new hypotheses about the nature of matter or the origin of DNA.[12]

ORCHARD RUN

WHAT IS AN *orchard run*? The expression refers to fruit as it is picked from the tree and shipped from the orchard—that is, before it is prettied up for the supermarket. Thus, orchard run apples may vary in size and color and may have marks from branches or hailstones; they are unpolished but tasty. My father, an orchardist, wrote a weekly column for the village newspaper. He called the column *Orchard Run* to reflect the freedom he intended to exercise in writing from his own perspective, ruminating on various topics from the use of pesticides to the most effective way to collect wild honey to the politics of the water management in our near-desert landscape.

For reasons paralleling those of my father, Wally Smith, I have used the notion of *Orchard Run* to allow myself to ruminate in a rather free-and-easy way on being comfortable with my brain—how I came to feel at home with the discoveries that teach me wonderful things about my brain's style of business. It gives me flexibility to muse in my mental orchard without worrying too much about coming across an imperfection or even an embarrassment here or there.

The story of getting accustomed to my brain starts long before I came to study the brain systematically. Like many children, I saw a grandparent gradually change from a competent, engaging, and loving human to someone who was fumbling and confused—disengaging, by increments, from the world and from me. My mother, who had trained as a nurse, was forthright in answering my questions, as best she could, given her own limited knowledge. Her explanation was all about the brain and about the decay of the brain and the loss of being. Grandma MacKenzie's memories, her ability to play the piano, knit, and grow vegetables all depend on her brain. Exactly *how*, of course, my mother could not say.

I knew a little about the brain; living on a farm in a rather isolated mountain valley in British Columbia, I could never be utterly detached from the reality of what was inside animals. I knew quite a lot about what was inside chickens, because I had to help clean them for meals. Inside I sometimes found eggs forming at various stages of development. I took apart the hearts, crops, and gizzards to see what was in their insides. It was the spinal cord we disrupted when we killed our hens by wringing their necks. I knew what bits of spinal cord looked like from examining the chicken necks; turkey necks were even better because they were bigger. I knew that when my cow Goldie was put down, they shot her in the head because that was certain death, and I knew from the butcher shop what the inside of a sheep's head looked like. All that loose and lumpy stuff on the tray behind the slanted glass was brain tissue. And I knew I, too, had a brain.

What will happen to Grandma when she dies? My mother, a practicing Christian, was very down to earth. Having nursed in the rural north and in hospitals with no antibiotics and little to control pain, she gave selected religious ideas her own particular emphasis, deftly putting aside disfavored ideas as "unworkable." On the matter of heaven, she would firmly opine, "It is more important to worry about the here and now. You can make a heaven or hell of your life right here." She emphasized that Grandma had lived a good life and that we all die sometime.

This was in keeping with the sorts of things my friends heard from their parents, so maybe it was just the no-nonsense way they adopted as part of surviving in a world where, as they ruefully reminded us, the wolf was never far from the door.

A farming village includes remarkable variability among its inhabitants, and my friends and I were curious about the so-called odd folks. Not hidden in attics (who had an attic?), the odd folks went about their business among the rest. Taught not to stare or be disrespectful, we nonetheless did discreetly inquire. We puzzled long and somewhat worriedly about Megan, who wore flowery dresses but even when she shaved regularly had a visible beard and mustache. Sitting behind her in church, I found this incongruity more gripping than the sermon. Megan's condition opened up questions about sexuality and personality and about what makes some of us keenly interested in boys and others not.

In the days before Down's syndrome could be identified in early pregnancy, quite a few children were born with that condition. Louise was my age, and her contrast with our classmates made us wonder what it was like to be unable to learn to read or milk the cow. Village life raised many puzzling questions: Why did Mrs. Herbert kill herself with rat poison on Easter morning while her husband and boys were in church? Why did Robbie Franklin constantly natter away to an unseen companion even though he was a grown-up? How was it that Mr. Fitzgerald's hands shook all the time?

My parents' approach to life implied that natural causes, poorly understood though they might be, were at the heart of the matter. Even suicide was attributed to what was then called "a nervous breakdown," which did not mean anything very specific except that something had gone amiss with the brain. Schizophrenia, likely responsible for Robbie Franklin's various delusions, was not called that, but his condition was regarded as a disease of his brain. It was a disease that the victim, who deserved our sympathy and certainly not our ridicule, could do little about. And Mr. Fitzgerald's trembling? My mother had seen many similar cases, and she sadly explained that the shaking was the result of a disease in the brain and that it would eventually kill him, unless, with luck, something else got him first.

I began to learn neuroscience in the mid-1970s after having begun a career in philosophy. This transition was motivated by the realization that if mental processes are actually processes of the brain, then you cannot understand the mind without understanding how the brain works. Studying the brain and thinking about how it works became a joyous obsession. Almost nothing about the brain, from tiny molecules shifting between neurons to the whole nervous system, failed to be fascinating. What is the *me* in all this—and, for that matter, the *we* in all this, my husband Paul and I wondered.

Because we are both philosophers, the big questions were very much in our minds at the end of the day, however many papers from neuroscience we devoured and explained to each other. To the exasperation of many of our philosophical colleagues, we pursued the idea that the nature of consciousness, language use, thought, and feelings would be illuminated by understanding the brain. To us, however, asking the ancient, traditional philosophical questions within the framework of neuroscience felt as natural as climbing a ladder to pick plums. We felt uncommonly lucky to be alive at a time when neuroscience was coming of age.

THE PRAGMATIC PERSPECTIVE

IS MAGIC more comforting than mechanism? Mostly not. The apparent comfort of ignorance quickly evaporates when something goes wrong in the nervous system. Faced with a degenerative disease, such as multiple sclerosis or senile dementia, we find that the mysteries, perhaps hitherto comforting, become instead a wretched obstacle to understanding and hence to possible intervention. Although you may feel that you have more control if the world is governed by magic than by mechanism, in fact the very opposite is true.

The relief that comes with insight into causation can be profound. I have a brother whose muscular development in childhood was more female than male and whose adolescence did not match up with the normal male milestones. At 25 he was finally diagnosed with Klinefelter's syndrome, meaning he had both the female sex chromosomes, XX, and also a male sex chromosome, Y. Thus, instead of the usual two sex chromosomes, he had three sex chromosomes: XXY. His relief, when the diagnosis finally came, was intense. At long last he could understand why he was different, why he had such poor impulse control and planned poorly, why he had so few whiskers but still liked girls. At first I was mildly surprised that he did not rail against cruel fate. But he absolutely did not. Instead, he found tremendous consolation in the physical explanation, along with release from the fear that somehow it was all owed to a character flaw for which he was wretchedly responsible. Most admirably, perhaps, it gave him a graceful determination to find harmonious ways of living with what he is. And remarkably, he has succeeded well in that.

Over the last 60 or so years, we have seen the control of diseases such as polio and whooping cough, the taming of other diseases such as diabetes and epilepsy, the prevention of conditions such as spina bifida and gangrene. The changes in my lifetime have been stunning. The watchmaker (repairman, really) in our village had only one leg; his trouser was pinned up and he used a crutch to get about. As I waited one day to collect my father's pocket watch, Cyril explained: 15 years before, he had been hiking far off in the hills, and distracted by a bobcat, he had trod on a rattlesnake, who then bit him on the leg. Knowing he needed to extract the venom, Cyril cut his leg open and tried to suck out the poison. The weather was hot, and he struggled to find his way back to treatment. After a day hiking out, the wound was infected, and by the time he reached the village, gangrene had set in. The leg was amputated. I had an intense respect for antibiotics as my two legs pedaled me home on my bicycle.

Although medical progress in our lifetime has been extraordinary, strikingly, the diseases of the nervous system have been largely intractable so far, with little available in the way of treatment. My closest childhood friend is afflicted with multiple sclerosis. Slowly, in the most cruel way, she has lost the use of her muscles, even those we count on for our dignity. Very little can be done even to slow the progress of the disease. She is stalwart, but as a nurse with a strong respect for reality, she knows where this is going. The cause remains unknown, and treatment is marginal.

A beautiful adolescent in my logic class went from being cheerful and clever to being delusional and disordered, as schizophrenia cinched its hold. She firmly believed a man-sized rabbit lived above her apartment and made terrible statements and thumping noises when she tried to sleep. Because neurological diseases bore into our very being, we have the most urgent need to understand them and figure out how we can stop them. When they strike, the brain's schemata for getting us around our physical and social worlds break into fragments like a shattered laptop.

Over the last century, attitudes toward neurological dysfunction have changed dramatically. Melancholia (what is now called chronic depression) or phobias were often surmised to be character flaws—flaws that could be overcome with sufficient gumption and, according to some opinions, hard work. And centuries before that, neurological dysfunction was often chalked up to the work of supernatural elements. Premenstrual syndrome was explained to me by the village dentist (of all people) as owed to sorrow at failing to become pregnant. As for those who had PMS but were greatly relieved *not* to be pregnant, the sadness was . . . well, still there but unacknowledged! (His may not have been a widespread view, and to be fair, he did have some other unusual convictions, all merrily expounded while the patient's mouth was full of tools. As the only dentist in the village, he and his odd opinions were tolerated in the office and gently debunked at home.) Autism was thought to be owed to "cold mothering" in some quarters. Obsessive hand washing, a mere 80 years ago, was widely assumed to be a manifestation of repressed sexuality, the ever-handy explanation for many behavioral oddities. Notoriously, some presumed aspect of toilet training was invoked to explain stammering, shyness, bed-wetting, chronic fibbing, insomnia, masturbation, having a tic, and being "boy-crazy."

It does seem generally true that as we come to understand that a particular problem, such as PMS or extreme shyness, has a biological basis, we find relief—relief that our own bad character is not, after all, the cause and relief because causality presents a possible chance for change. If we are lucky and current science has moved along to understand some of the causal details, interventions to ameliorate may emerge. Even if a medical intervention is not available, sometimes just knowing the biological nature of the condition permits us to work around, or work with, what cannot be fixed. For some problems, such as bipolar disorder and chronic depression, medical progress has been greater than for other problems, such as schizophrenia and the various forms of dementia. As more is unraveled about the complex details of these conditions, effective interventions will likely be found. The slow dawning of deep ideas about the brain and the causes of neurological dysfunction has lifted from us the cruel labeling of demonic possession or witchery.

This book is structured around the issues that tend to give us pause as we contemplate what understanding the brain might signify. Given some leeway, each issue acts rather like the center of a force field, drawing in to itself its own particular assortment of data, stories, and reflections. Special sensitivity halos our sense of self, our sense of control, our take on moral values, consciousness, sleep, and dreaming. We feel less sensitivity toward how our brain controls our body's temperature than how it manages our autobiographical memory, though management of body temperature is certainly a crucial function. The sensitivity is spawned partly because functions such as self-control and consciousness are at the core of our very being. It is owed partly and reasonably to fear of the unknown, fear of changing our worldview and our self-view. It is owed to uncertainty about what exactly will change and how.

My take on the roster of sensitive issues is that although much is still unknown about the nervous system and how it works, what *is* known begins to free us from the leaden shackles of ignorance. It makes us less vulnerable to flimflam and to false trails. It grounds us in what makes sense rather than in the futility of wishful thinking. It adds to the meaningfulness of life by enhancing the connections between our everyday lives and the science of how things are. Harmony and balance in our lives are deepened and enhanced by that connectedness.

I began by saying that my brain, not a soul, holds the key to what makes me the way I am. That assertion is backed by science—actually a lot of science and not only neuroscience. Still, a fair question is this: Does the idea of the soul really deserve to be shelved, much as we have shelved the idea of spontaneous generation of mice from dirt or the idea of geocentrism or animal spirits? Might we have a soul *as well as* a brain? The next chapter takes a closer look at the heyday and then the slump of the hypothesis that the soul is what gives us a mental life.

Chapter 2

Soul Searching

For vast stretches of human history, no one knew that it is the brain that allows us to walk, see, sleep, and find mates and food. We just did those things—walked, saw, slept, found mates and food. You do not need to know that you have a brain for the brain to operate very efficiently in getting you around the planet and seeing to your well-being. You do not have to stoke up and direct your brain; it stokes up on its own and directs you.

The human brain has been shaped by hundreds of millions of years of evolution. A powerful driver in the evolution of the brain was the importance of moving the body and making predictions so as to guide movement appropriately. An animal needs to move its body for all the necessities—to get food and water, avoid predators, and find a mate. For an animal to thrive, its brain needs to respond to pain or cold, thirst or lust, in appropriate ways, and those ways typically involve organizing bodily movement. Better predictions allow a brain to make more successful movements, and more successful movements increase the animal's chances of surviving and reproducing, thereby spreading the genes that built that better brain. To perform these jobs more efficiently and thus to compete more success-fully in a tough world, complex brains evolved neuronal circuitry to model the body—its limbs, muscles, and innards—along with relevant aspects of the outside world.[1]

First, consider the brain circuitry organized to generate a neural model of the world outside the brain. Processes in this neural organization model events in roughly the same way that the features of a map model the features of the environment. The typical map does not *resemble* all aspects of those features—the squiggles for the river are not actually wet, for example, nor is the real river only a millimeter wide and com-pletely dry. Still, the map does constitute a *representation* of certain relevant aspects of the environment. In particular, map features have the same spatial relationships with each other as geographical features in the world do. That is what makes the map a faithful model, and that is what makes the map useful for navigation. Thus, in both the map and the real world, the river's headwater is closer to the mountain than to the sea, the river makes a bend to the north before going into the sea, there is a broadening of the river as its descends from the mountain, and so forth. In a somewhat similar way, a satellite camera's image of Earth represents Earth, including the color differ-ences between oceans and land. That image is not literally wet or cloudy, but it represents the oceans and the clouds.

Caution: Before getting too cozy with the map analogy, let me be clear about where it breaks down. When I consult a road map, there is the map in my hand and, quite separately, there is me. The map in my hand and I are not one. In the case of the brain, there is just the brain—there is no separate thing, me, existing apart from my brain.[2] My brain does what brains do; there is no separate *me* that reads my brain's maps. This disanal-ogy with using a road map is part of what makes understanding the brain so difficult, especially since the idea of someone in my head reading the brain's maps keeps sneaking back into my thinking. Ironically, however, that very disanalogy contributes to what makes neuroscience so thrilling. We understand, more or less, how I can read a road map. We do not understand nearly as well how I can be smart because my brain maps my inner and outer worlds, absent a separate *me* to read those maps. I want to know how all that world mapping and me mapping is done.

With the caution and thrill duly registered, we can now return to the idea of brain models, agreeing that no separate person is in our brain reading the maps or using the models. The brain models aspects of the external world, thereby creating the informational structure that allows us to interact productively with things in the external world. This roughly means that by virtue of brain organization, there is a relationship between the external events and particular brain activities that enables the brain to navigate the world in order to find what the animal needs to survive. When sensory representations in the brain are connected in the right way with the motor system, successful behavior, such as self-maintenance, is the outcome. Thus, the animal flees a predator or lands a prey or deposits some sperm, for example.

The "designer" of brain organization is not a human cartographer, but biological evolution. If an animal's brain misrepresents its domain, mistaking a rattlesnake for a stick (rather like mistaking 4th Avenue for Stanton's Creek), or if the mapping functions do not direct the motor system to produce the appropriate behavior—approaching a predator when it should flee—the animal will probably be killed before it has a chance to reproduce.

Suppose you hear the sound of a crow cawing. Although the activities of the auditory system neurons do not literally resemble sound waves, those activities zero in on the right place in your learned map of the typical sounds in the world, such as the cawing of the crow. The various differences in the physical patterns of sound waves made by whistles or cries or bangs are represented in your neural model as differences in the positions within your internal neural map.[3]

As we look at the specializations of the brains of different species, it is evident that brains have evolved to map their worlds according to their sensory and motor equipment and its evolution. More accurately, we see a *coevolution* of brain equipment and body equipment. Bats have an exceptionally large part of their cortex devoted to processing auditory signals because they use a sonarlike system to find, identify, and catch objects at night. Monkeys and humans, who tend to sleep at night, have exceptionally large cortical areas engaged in processing visual signals.

For their body size, rats have a huge somatosensory cortex, much of it devoted to mapping the activity of their whiskers. This is owed to the fact that rats do much of their business in the dark, where vision is useless but where smell and touch have proved invaluable. Rats and other rodents use their whiskers (vibrissae) to get information about the size of holes and crannies, to palpate objects, and to orient themselves in their surroundings. Their whiskers rhythmically sweep back and forth between 5 and 25 times per second (known as *whisking*), while the brain integrates the signals over time so that the rat can identify an object.[4] This is the rat equivalent of what we do when we scan a scene by making many eye movements across the terrain. Humans use their eyes but not their whiskers to navigate their external world, making saccadic eye movements about three times per second. Inevitably, I wonder what identifying another person by whisking instead of by visual scanning is like. A congenitally blind person who reads Braille probably has a pretty good idea of what whisking is like.[5]

I emphasize that the brains of animals map *their* worlds because no brain maps everything that is going on in *the* world as a whole. A brain maps what is relevant to how that animal makes its living. Accordingly, the brain models are beholden to the animal's sensory and motor equipment, as well as to its needs and drives and cognitive style. Thus, some birds have systems that are sensitive to Earth's magnetic field, guiding them at night on their migrations; some fish species—for example, sharks—have receptors to detect electric fields, allowing them to avoid electric eels, which can generate an electric shock to stun and kill their prey. Human brains do not map magnetic or electric fields. Luckily for us, however, we have other means, largely visual, for mapping our spatial worlds. Moreover, our flexible, problem-solving brains eventually made a device—a compass—that can give us information about Earth's magnetic field by virtue of the magnetic needle's causal connection to Earth's magnetic field and our mapping of the relation between the compass needle and north.

Moreover, in the brains of different species, the *resolution* of the map—the degree of detail—varies, too.[6] Here is what I mean: I could sketch you a crude map of the Yukon River on a table napkin, or you could buy a highly detailed map of the river and its environs. In a roughly comparable way, relatively simple brains have rather crude maps; more complex brains can map the environment to a greater degree of resolution.

Bats, for example, have very high-resolution maps of the auditory environment, because as mentioned earlier, they use an echolocation (sonar) system for detecting and identifying objects. (Many blind people use a similar strategy to navigate, making a noise by clicking their tongue against the roof of the mouth.) Notice, too, that the commercial map of the Yukon River could also contain representations of elevation, a rather *abstract* feature, without the map literally having little hills and valleys. Brains, too, can map abstract features without having neuronal hills and valleys. Importantly, they can map abstract *causal* relationships. There are simple causal relationships, such as between a splash in the stream and the presence of trout, or more complex ones, such as between the phases of the moon and the rise and fall of the tides or between something unobservable with the naked eye, such as a virus, and a disease such as small-pox. This latter kind of causal knowledge requires a cultural context that accumulates layers upon layers of causal knowledge, won through experience and passed on through generations.

Sophisticated social behavior can also be mapped, as in this example of a foraging crow. The crow has observed a husky get his kibble at dinnertime. One day the crow approaches the husky as he begins to eat. The crow stealthily glides down behind the dog and yanks on the dog's tail. The dog turns, the crow flies, tantalizingly low but just outside the margin of danger, out of the yard and down the street. The excited dog gives chase. After a few minutes of chase, the clever crow flies directly back to the food dish and helps himself.[7] The behavior strongly suggests that the crow predicts that his yanking the tail and his flying low will cause the dog to give chase. Thus is the dog lured away from his food.

Brain circuitry also supports a neural model of the *inner* world. It maps the muscles, skin, gut, and so on. By this means, you know the position of your limbs or whether something nasty has happened to the skin on your nose or whether you need to vomit. We are so accustomed to the smooth operation of our brains that we take it for granted that knowing where our legs are is just obvious and simple, even from the point of view of the brain. Not so. Not at all. Sometimes the mapping functions can get disrupted as a result of injury or disease. Then the complexity of brain mapping is revealed, as we shall now see.

A favorite way to spend time after our chores were done was to take our bicycles out on the back roads of the hills above the valley. For miles and miles we could explore, seeing no one. The roads were dirt, and if your bike plowed into a patch of gravel, you could lose control. My friend Christine and I peddled hard to the top of a hill, then flew fast down the hill to the bridge over the creek below. We both spun out. I bowled over into the creek and lost a fair bit of skin on my bare legs. Christine hit her head.

Only 12, I had never seen the symptoms of serious concussion, but after a few minutes it was clear that something was wrong with Christine's head, in addition to an egg-sized lump above her right ear. She did not know where she was or how she got there. Sitting at the creek edge, she stared blankly at her left leg and asked whose it was. She asked the same question about every 30 seconds. How could she not know the leg was hers? Who else could it belong to? I finally asked her. She said, "Maybe a tramp." A semireasonable answer, inasmuch as tramps were not uncommon in the area. Clearly, I could not let her ride home. Just as clearly, I could not leave her to get help.

The story ends well because about an hour after the mishap, a logging truck rumbled down the road. I flagged the driver down, we tucked the bikes in among the pine logs, and we got her home. After a few days of rest, Christine was fine, as the doctor had calmly predicted. She knew her leg was hers and was thunderstruck when told there was a bit of time when she did not. She remembered essentially nothing of the whole episode.

Later I came to know that subjects with damage to the parietal cortex of the right hemisphere may believe that limbs on the left side, such as an arm, do not belong to them, a condition known as *somatoparaphrenia*. It is often accompanied by loss of movement and loss of feeling in those limbs. Otherwise clearheaded, these patients say very bizarre things about the affected limbs. For example, one patient whose arm was paralyzed said that she could indeed move her arm. When she was asked to point to her nose and could not, she nonetheless insisted that yes, she had in fact pointed to her nose.[8] Another said of her left arm that it belonged to her brother.

Neuroscientist Gabriella Bottini and her colleagues reported a remarkable case in 2002 of a woman who had suffered a stroke in her right hemisphere, with typical loss of mobility in her left arm. One effect was that she believed firmly that her left arm belonged to her niece. She also seemed unaware of being touched on that arm, a not unusual feature in such a case. In one assessment, however, the doctor explained to her that he would first touch her right hand, then her left hand, and then her niece's hand (when he actually touched her left hand). He did this, asking her to report what she felt each time. She felt the touch to the right arm, felt nothing on the left, but surprisingly did indeed feel the touch to the left when the doctor touched what she believed was her "niece's hand."[9] The patient agreed that it was odd that she felt touches in her niece's hand, but was not especially perturbed by the oddity.

I am dwelling on somatoparaphrenia delusions because they truly push our strongest intuitions about body knowledge, reminding us that intuitions are only intuitions. They are not always reliable, and they are no guarantee of truth. Knowing that your legs are *yours* or that a feeling on your legs is *your feeling* seems dead obvious. Because such knowledge typically is not something you consciously figure out, philosophers such as Ludwig Wittgenstein were motivated to assume that it was impossible for you *ever* to be wrong about whose leg this is. Not just irregular or unusual—but flat-out impossible. The problem was that he was listening *only* to his intuitions, which seemed so deeply true. He was not letting the data tell him what is and is not possible.

Normally, you are not wrong about your leg or arm; but in fact, your brain has to be working in just the right way, below the level of consciousness, for you to know that the legs you see are indeed *your* legs and that the feeling on your leg is *your feeling*. Disruption of processing resulting from brain damage, especially damage to the parietal area of the right hemisphere, means that sometimes we *are* wrong.[10] The arm in the bed did not belong to the patient's niece, however strong the patient's intuition that it did.

In addition to the brain's modeling the body it inhabits, some parts of the brain keep track of what other parts of the brain are doing. That is, some neural circuits model and monitor the activities of other parts of the brain. For example, when you learn a skill, such as how to ride a bicycle, subcortical structures (the basal ganglia) get copies of your current goal along with copies of the current motor commands from the cortex. When you get a movement right, given your goal, neurons in the basal ganglia in effect say "nailed it!" by releasing the neurochemical dopamine. The result of the precisely timed dopamine release is that connectivity changes are made in various parts of the brain to stabilize the circuitry supporting the set of movements that were right. The next time you try to ride, those movements—the right movements—will more likely be generated by your motor cortex. If the basal ganglia fail to get a copy of the current goal or fail to get a copy of the motor signal, or if the precise timing of the signal relay is messed up, the brain cannot learn. That is because it has no way to know which, among the many movements commanded, was the winning one—the one that caused the right movement.[11]

Here is another example of the brain monitoring the brain that actually changes our visual perception. Imagine that you hear a sudden bang and turn your head to see the source. You locate the source as a pot that fell off a shelf. In response to the bang, neurons in the motor cortex made a decision to turn the head in the direction of the sound. Patterns of light smeared across your retina as your head turned. In addition, a *copy* of the head movement signal went to other areas of the brain, including your visual cortex. This movement signal copy (*efference copy*) is very useful because it tells your brain that *your head*, not something in the world, is moving. Absent an efference copy, your visual system would represent the shifting patterns on the retina as owed to things moving *out there*. Then confusion would reign.

This organization for efference copy is very clever, because it means that you are not misled about the origin of the shifting patterns on your retina when you move your head or when you move your eyes or your whole body. It is, I suspect, an important source of data that the brain uses in generating the complex sense of *me* versus *not-me*. Most likely, you are not even visually aware of moving patterns of light when you move your head. Your brain is extremely good at downplaying awareness of those moving patterns because those retinal movements are inconsequential so far as interpreting the external world is concerned.

Sometimes your brain can be fooled. Suppose you are stopped at a red light and the car next to you rolls back unexpectedly. Out of the corner of your eye, the movement is picked up. For the first second, you are apt to think that *you* are rolling *forward*, as that would be most probable, given the situation. With additional input, however, the brain makes the correction. Improbably, the car next to you is rolling backward. Is any of this soul business? No. This is brain biology, glorious efficient biology, doing its magnificent job, but not, of course, flawlessly.

A neuroscientist friend was fascinated by efference copy, wondering what his visual experience would be like if he paralyzed his eye muscles (transiently, by means of an injected drug), then formed the intention "eyes, move right." In this condition, a copy of the intention to move would be sent to the visual system, which would not know of the eye muscle paralysis. Many years ago, he carried out the experiment on himself. What did he experience?[12]

His visual experience was that the whole world made a jump to the right. Essentially, the brain interpreted its visual input from the retina on the assumption that eye movement actually occurred, since, after all, the intention "eyes, move right" was registered. Since the visual input had not changed, however, the brain concluded that the world must have moved. Smart brain, to light on a reasonable guess.

This was a heroic and rather risky experiment, never published because it was definitely under the radar. Moreover, the experimenter was the subject and the *only* subject. It did nonetheless capture my imagination as I pondered the striking effect on visual perception *itself*—what you literally *see*—when you suppress the movement of the eyes. The brain evidently counts on the tight coupling of the *intention* to make an eye movement and the eye movement actually occurring. A breakdown, via the experimental paralysis, of that coupling affected the organization for distinguishing *my movement* from *external movement*.

Reflecting on efference copy made me appreciate anew that a basic job for a brain is to distinguish the *me world* from the *not-me world*. Efference copy is probably only one trick, albeit an important one, among many for achieving that distinction between *me* and *not-me*.

To return to the main theme of *representations and maps*, notice that when some part of the brain "reports" on its state to another part, you experience these reports as feelings, thoughts, perceptions, or emotions. You do not experience them in terms of neurons, synapses, and neurotransmitters. Similarly, when I am thinking about going fishing tomorrow, I am not directly aware of my thinking *as* a brain activity. I am aware of visual and movement images, maybe accompanied by a silent monologue. It goes sort of like this: I imagine myself at the creek side, night crawlers tucked in a can and speckled trout spooning in the cool shadows. I am not aware of this as the brain working. I am aware of it simply as *making a plan*. I do not have to tell my brain how to do any of this. The brain's business is to do it. When I feel hungry, I am not aware that my brain makes that feeling; when I feel sleepy, I just feel sleepy. My brainstem, however, is busy making me feel sleepy. Among other things, it is decreasing the level of neuromodulators, in particular norepinephrine and serotonin.

Why doesn't the brain make it self-evident that it is doing all these things? "Oh, by the way, it is me, Brainsy—in here in your head. I am what allows you to maintain balance and chew your food; I am the reason you fall asleep or fall in love." Nothing in the ancient environment of brain evolution would select for brains that could reveal themselves thus. Similarly, there is nothing in our current environment to select for kidneys or livers that announce *their* existence and modus operandi. They just work the way they evolved to work.

By contrast, advantages do accrue to animals whose brains can map and remember the spatial layout of their neighborhood and its food sources. Consequently, many animals have nervous systems that are remarkably good at spatial learning. They know where home is, where food is cached, and where predators lurk. If you eat fruit, it is advantageous to have color vision so you can distinguish ripe from unripe fruit. If you are an owl hunting mice in the dark, it is advantageous to have superb sound location. If you are a social mammal or bird, it is useful in predicting the behavior of others to see their behavior in terms of having goals and feelings.

Because the models that brains deploy do not on their own reveal the nature of the underlying brain mechanisms, coming to understand the brain has been exceedingly difficult. When I first held a human brain between my two palms, I muttered to myself: "Is this really the sort of thing that makes me *me*? How can that be?"

BODY AND SOUL

EXACTLY WHEN humans first began to understand the importance of the brain as the substrate for thinking and behaving is unclear. Certainly, there was nothing like a date in antiquity when the fact was established and thereafter widely believed. By contrast, there was a time in antiquity when it was discovered that by adding carbon to molten iron you could make it wonderfully strong—*steel*. Still, observations in prehistoric times regarding the dire effects of severe head injuries resulting from battle or accidents must have provoked a general appreciation of the importance of protecting the head from severe insult.

It is known that the great Greek physician Hippocrates (460–377 BCE), contemplating such data, opined that the brain is the basis for all our thoughts, feelings, and ideas. Precisely how he came to that pioneering insight in such ancient times is not known. As a physician, he undoubtedly performed dissections on people who died after stroke, and he may have seen soldiers with localized head wounds that correlated with the loss of specific functions such as vision or speech. He probably saw difficult births that left babies with severe disabilities. Hippocrates, like other ancient Greek thinkers, was a naturalist, not a supernaturalist; in looking for explanations of how things worked, he sought his explanations in the natural world. In his down-to-earth way, he reckoned spirits and gods and otherworldly stuff to be rather sterile in the explanatory business. Plato (428–348 BCE), by contrast, had a mystical bent. He assumed that each of us is endowed with a soul that lives before birth, inhabits the body in life, and departs after the body's death, dwelling blissfully in a Soul Land, which also contains all absolute truths. It was sort of like believing that the trash can displayed on your computer screen's dock survives the destruction of your computer—it goes to Virtual Reality Trash Can Land. Plato's musing launched, at least in the Western tradition, the idea of a nonphysical soul—a dualism of stuff. Even earlier, Hindu philosophers had come to a similar conclusion.

On Plato's theory of the mind, understanding and reason are the business of soul stuff, while movement, the basics of perception, and so forth, are the business of physical stuff—the body. Genuine knowledge is arrived at through reflection, according to Plato, and is achieved bit by bit, despite the unfortunate interference of the physical body in the soul's noble attempt to remember the absolute truths to which it was privy while resident in Soul Land.

Aristotle (384–322 BCE), despite being Plato's prize pupil, was more firmly planted in the physical world. Like Hippocrates, Aristotle favored naturalism. He looked mainly to the organization of matter to explain how things work. The son of a physician, Aristotle was used to thinking about the body and mind in a medical framework. Although Aristotle's ideas about psychological states are complex and susceptible to interpretation, he clearly thought that all emotional states (anger, fear, joy, pity, love, hate) are actually states of the body. He was less clear about whether the intellect—when it is doing mathematics, for example—is also a bodily function. Suffice it to say that Aristotle was very sophisticated about biological matters and also sensitive to the complexity of human reasoning. As he humbly says, "Grasping anything trustworthy concerning the soul is completely and in every way among the most difficult of affairs."[13]

From Plato the mystic and Aristotle the naturalist emerged the two Western traditions: dualism (soul stuff *and* brain stuff) and naturalism (only brain stuff). Some 300 years later, the Christian Era (also called the Common Era) began. During the early part of the Christian Era, a prominent idea was that after a Christian died, the body would be resurrected and would physically ascend into a region somewhere above the moon. This did not require the idea of a Platonic soul, just the body.

Not surprisingly, many questions arose regarding the details of the promised resurrection and afterlife. People wondered at what age their body would be resurrected (in their prime, in childhood, or when decrepit?), whether a long-amputated limb would be reattached, whether wounds would be healed or remain festering, whether second or first husbands would be *the* husband (or whether people would have spouses at all), and so on. Eternity is a long time, much longer than a lifetime, so these questions were not trivial or merely academic.

Obviously, there were inconsistencies in the idea of physical resurrection, since decomposition of the body after death was well known. One way to square resurrection with the corruptibility of the human body was to argue that in heaven, Christ changes our corruptible bodies into spiritual, imperishable bodies.[14] This borrowed a bit from Plato's idea that upon death the soul returns to Soul Land, but at the same time, it was meant to reflect the Christian belief that Jesus changed everything when he arose bodily into heaven. Better bodies—glorified bodies—are the heavenly reward for believing in Jesus. The idea of a spiritual body may seem a bit like the idea of a square circle, and the details of how exactly this worked were left conveniently vague.

Much later, in the seventeenth century, René Descartes (1596–1650) puzzled long and thoughtfully about the nature of the mind.[15] He knew that the brain was important, but believed its role was essentially limited to two basic functions: (1) executing movement commanded by the soul, and (2) responding to external stimuli, such as a touch to the skin or the light entering the eye. He sided with the Platonic ideas of the soul as the key to explaining how it is that humans use language and can make choices based on reasons, achievements that he thought were absolutely beyond any physical mechanism. Why was Descartes's imagination so limited? Well, this was the seventeenth century, and the fanciest physical mechanisms with which he was familiar were mainly clocks and fountains. Although these devices could be impressive, they completely lacked scope for novelty. Human minds, by contrast, were capable of impressive novelty, especially in speech. Had Descartes had the opportunity to use my MacBook Pro, he might well have stretched his imagination much further.

In any case, Descartes concluded that all mental functions—perceiving, thinking, hoping, deciding, dreaming, feeling—all are the work of the nonphysical soul and *not* the brain. Where did he suppose the handoff of information between brain and soul takes place? In the pineal gland, providentially located in the middle of the head. As it turns out, the pineal gland's main function is to produce melatonin, which regulates sleep/wake functions. Trafficking signals between body and soul is not one of its jobs after all. Descartes did not get this wrong because he was dim-witted. On the contrary, he was unusually brilliant, especially in geometry. He got it wrong because so very little was known about the brain at the time he lived.

By the nineteenth century, a few scientists, but especially Hermann von Helmholtz, realized that souls, special energies, occult forces, and other nonphysical things were likely a dead end so far as explanations of mental functions such as perception, thinking, and feeling were concerned. With great insight, Helmholtz proposed that many brain operations happen without conscious awareness. He came to this hypothesis while pondering the fact that when you look around, you can see and size up a complex visual scene in less than half a second (500 milliseconds), all without any conscious thinking. Sizing up a scene is very complicated, since the only thing that stimulates your retina are patterns of light. Yet you see colors, shapes, motion, relative position in space, and you instantly recognize familiar faces and other objects. So how does the brain get from patterns of light to "Hey, that's Queen Elizabeth!"?

Helmholtz reasoned that by the time you see and recognize a familiar face, a lot of nonconscious processing has already been done, and done with remarkable speed and amazing accuracy. He did not have a clue about the exact nature of that processing, since almost nothing was known about neuronal functions. Nevertheless, that such processing occurs, in parallel pathways and below the level of consciousness, is entirely correct. (I am using the words *nonconscious* and *unconscious* interchangeably in this context. See Chapter 8 for a fuller discussion of the scope of nonconscious brain functions.)

So Helmholtz rightly realized that the brain has to be doing lots of processing that is nonconscious and that to understand such processing, paying attention to *conscious* activities is not enough. Moreover, if conscious and nonconscious processing are interdependent, identifying a nonphysical soul with only conscious activity is far-fetched. It would be like identifying someone's nose as the whole body.

By the mid-twentieth century, the steam had largely gone out of the dualist hypothesis as an account of thoughts, perceptions, and decisions. It was not so much that there was a single experiment that decisively showed that the brain does mental jobs such as seeing and deciding. Rather, it was the accumulation of evidence, from every level of research on nervous systems, from neurochemicals to whole systems, that collectively detracted from the idea of ghosty souls. This is typical of science generally, where a well-entrenched paradigm rarely shifts overnight, but imperceptibly, bit by bit, as evidence accumulates and minds slowly reshape themselves to the weight of evidence. In specifically religious contexts, however, dualism of some vague sort continued to be appealing.

Evidence did accumulate from many different directions. For example, physical changes in the brain produced changes in supposedly soul functions, such as consciousness, thought, and reasoning. Inhaling an anesthetic such as ether caused people to lose consciousness; ingesting a substance such as mescaline or peyote caused people to experience vivid hallucinations. Neurologists reported very specific losses of function correlated with damage to particular brain areas. A person who suffers a stroke in a very specific place in the cortex (the fusiform) will likely lose the capacity to recognize a familiar face; a stroke in a somewhat different area will cause the loss of the ability to understand speech. Loss of social inhibition may follow a stroke that destroys the prefrontal cortex just behind the forehead. All these phenomena seem to point to the nervous system, not to nonphysical, spooky stuff.

One discovery in particular did cause a brouhaha among the remaining die-hard dualists. In the 1960s, Roger Sperry and his colleagues at Cal Tech studied patients whose hemispheres had been surgically separated as a last resort for the control of debilitating epileptic seizures. These subjects became known as the split-brain patients. Careful experiments showed that when the nerve bundle connecting the brain's two hemispheres is surgically cut, the patient's two hemispheres become somewhat independent cognitively. Lower structures, such as those in the thalamus and brainstem, are not separated—hence the qualification "*somewhat* independent."

In split-brain subjects, each hemisphere may separately experience the stimuli delivered exclusively to it. If, for example, a key is placed in the left hand and a ring is placed in the right hand and the subject is asked to use his hands to point to a picture of what he felt, the left hand points to a picture of a key and right hand to a picture of a ring.[16] A split-brain subject may even make opposing movements with the two hands—the left hand picking up the phone, the right hand putting it down. Or if a visual stimulus, for example, is presented to just one hemisphere, the other hemisphere knows nothing about it. This was a completely stunning result. Did splitting the brain split the soul? The soul was supposed to be indivisible, not divisible like a walnut. But there they were, the split-brain results, available for all to see: if the brain's hemispheres are disconnected, mental states are disconnected. Those results were a powerful support for the hypothesis that mental states are in fact states of the physical brain itself, not states of a nonphysical soul.[17]

Descartes's notion of the soul did not work out very well with physics either. The problem is this: if a nonphysical soul causes events to happen in a physical body, or vice versa, then the law of conservation of mass energy is violated. The trouble is, so far anyhow, that law seems very resilient against all comers. Well, maybe—just maybe—it does happen. But how? Even very roughly, *how*? How can energy be transferred from a completely nonphysical thing to a physical thing? Where does the soul get its oomph to have such an effect? What kind of energy does a soul have? Is it measurable? If not, why not? Descartes, interestingly, was fully aware of *that* particular problem and despaired of ever solving it.

Once you give slow thought to what sort of thing a nonphysical soul might actually be, awkward facts begin to pummel the idea's plausibility. For example, think about what happens when my dentist "freezes" a nerve in my wisdom tooth and my "soul" ceases to feel pain in that tooth. The neuroscientist has a well-established explanation regarding why I cease to feel pain. Substances such as procaine (trade name Novocain) injected close to neurons emerging from the tooth shut down the neuron's capacity to respond. The result is that no pain signals from the neuron are sent to the brain. Moreover, we know exactly how procaine does that. For a neuron to be active, sodium ions first get pumped out of the cell, and then when the neuron gets a stimulus, the sodium channel opens and the ions flood back into the neuron. Procaine temporarily blocks sodium channels and hence prevents the neuron from signaling. With time, the procaine denatures and hence the effect disappears. The neuron's capacity to respond returns and the painful feeling returns.

The explanation of how procaine blocks the transmission of pain signals is satisfying because it provides details of mechanism, it can easily be tested, and the details fit with what else we know experimentally about pain and neurons. Such "fitting with the rest of the body of knowledge" is called *consilience*: the greater the consilience, the greater the coherence and integration of phenomena and facts. Note, however, that consilience is not a *guarantee* that the explanation is right, because you can have a totally wrong theory whose bits and pieces happen to cohere.

This happened to Newton's contention that space is absolute, like an empty vessel, and is everywhere the same. This theory made sense of a huge amount of evidence. Einstein, however, conjectured that Newton's supposition about absolute space might be false and that mass might distort space. When Einstein's prediction was tested and found to be correct, Newton's theory had to be abandoned, though it had stood as certain for some 300 years. Space is *not* everywhere the same; huge gravitational bodies, such as the sun, measurably alter the geometry of space. Einstein's theory ended up having much greater explanatory power than Newton's and much more consilience with developing physics.

Back to my wisdom tooth. Can the dualist match neuroscience's level of explanatory consilience regarding why procaine blocks pain? Not even close. A dualist could say, well, the procaine also acts on the soul. But how, even roughly, does that work? What does it *do* to the soul—especially if procaine is physical and the soul completely not physical? This dualist proposal says nothing at all about mechanism. Consider the contrast with the neuronal explanation, which is all about mechanism.

In principle, a dualist could experimentally work out the details of a soul theory, finding out how souls work and what their properties are. Hypotheses could be tested. Experiments could be run. In principle, there could be a natural science of the soul that would explain why souls lose consciousness when the body inhales ether or why souls hallucinate when the body ingests LSD. In practice, however, there is no science of the soul. Apart from flimsy contrasts with the body (such as "the soul is *not* physical," "the soul has *no* mass or charge," "the soul has *no* temperature"), there has been no advance since Descartes's 350-year-old hypothesis. The odd thing is that dualists, even deeply convinced dualists, are not even trying to develop soul science, as though merely saying "the soul does it" for every *it* is explanation enough. It is not nearly enough.

We cannot be certain that no distinct soul science will ever flower, but as things stand, brain science seems to have the leg up on soul science. This suggests that soul theory is floundering because there is no soul. If you had to place a big-money bet, on which hypothesis would you put your money?

WHY IS IT SO HARD TO FIGURE OUT HOW THE BRAIN WORKS?

THE BRAIN is not an easy organ to approach experimentally. For one thing, it doesn't look like anything familiar—not a pump (as is the heart), nor a filter (as is the kidney). A neuron (the signaling cell in the brain and spinal cord) is *very* tiny, not visible with the naked eye. A bundle of neurons, such as those that make up the sciatic nerve, for example, is visible, but these bundles are composed of thousands of neurons. In the cortex, a cubic millimeter of tissue contains tens of thousands of neurons, a billion connectivity sites (synapses), and about 4 kilometers of connections.

Neurons cannot be seen as individual cells without a light microscope,[18] a device not extensively used in research until about 1650. Even then, special chemical stains had to be discovered to make one single very tiny neuron stand out against the rest of the millions of very tiny neurons closely packed together. Only thus could the basic structure of the neuron— an input end for receiving signals and a long connecting cable for sending signals—be seen. Techniques for isolating *living* neurons to explore their *function* did not appear until well into the twentieth century.

Making progress on how the brain works depended on understanding electricity. This is because what makes brain cells special is their capacity to signal one another by causing fast but tiny changes in each other's electrical state. So if you did not know anything about electricity, you would be stumped when you wondered how neurons can send signals and what a signal *is*. You might have thought that neurons communicated by magical forces. And for a long time, people *were* stumped, even after something was known about the basic structure of neurons.

Owing to Luigi Galvani's observations (1762) that a spark would make a frog's detached muscle twitch, the idea that electricity might be important for the function of nerves and muscles was on the table. But how did that work? Galvani himself did not understand at all what the relation was, largely because electricity was so poorly understood at that time. He conjectured that there was a special electrical bio-fluid that was carried to the nerves and muscles. Not until the early part of the twentieth century was it discovered that signaling of neurons depends on ions (charged atoms) abruptly moving back and forth across a neuron's membrane. Exactly how that produces a signal was finally explained in 1952 by two British physiologists, Alan Lloyd Hodgkin and Andrew Huxley. Their discovery revolutionized the science of the brain. But notice how recent that discovery was: 1952. After I was born.

Simplified, this is what Hodgkin and Huxley discovered. Like all cells, nerve cells (neurons) have an outer membrane, partly constituted by fat molecules, with special protein gates that open and close to allow particular molecules to pass in or out of the cell. In a resting neuron, the inside of the membrane is negatively charged relative to the outside, owing to active pumping out of positive ions, such as sodium. Negative ions, such as chloride, are sequestered inside. This voltage difference can change abruptly when the neuron is stimulated. This fast change in voltage across the membrane is what makes neurons special. For example, when you touch a hot stove, a heat-sensitive neuron responds, which means that sodium ions rush into the cell, briefly reversing the voltage across the membrane. Immediately thereafter, sodium ions are pumped back out, restoring the original condition. This fast reversal and restoration in voltage across the membrane is what is referred to as a *spike*. By placing a wire close by the neuron and attaching it to an amplifier, you can hear a snap when the neuron spikes.

Once started, the voltage change moves down the membrane of the neuron to the end (spike propagation). Also called a *nerve impulse*, this signal eventually (in a few milliseconds) arrives at the neuron's tail. This can trigger release of a chemical (neurotransmitter) that then drifts across a tiny space to the next neuron, docking at a special site and causing a voltage change in the receiving neuron. Alternately, if the neuron contacts a muscle cell, the muscle will then react—for example, by contracting. Oversimplified to be sure, this account opens the door to the world of nervous systems in all their complex glory.[19]

Accustomed as we are to all manner of electrical devices, it takes us aback to realize that as late as 1800, electricity was not understood. Many considered the phenomenon to be occult, never to be explained as a physical phenomenon at all. At the dawn of the nineteenth century, all that changed when electricity was clearly understood as an entirely physical phenomenon, behaving according to well-defined laws and capable of being harnessed for practical purposes.[20] Some people whined that this discovery was taking the divine mystery out of electrical phenomena. Others started inventing electrical devices.

Is there room among all the remaining puzzles in neuroscience for a soul (the nonphysical, Platonic variety)? Possibly. But not, it seems to me, remotely likely. Nevertheless, what has struck many people with a fondness for dualism is this: when a neuron responds with a spike, that neuronal response seems entirely different from the pain I feel when I touch a hot stove. How does the activity of neurons—possibly many, *many* neurons—produce pain or sounds or sights?

The answer is not yet known. There are many strategies for making progress in answering that question, however, some of which converge. So there is progress, if no fully detailed account of mechanism. (For more on this, see Chapter 9.) For example, much research has gone into trying to understand exactly what happens when a person is put under anesthesia and loses conscious awareness. Many anesthetics work by shutting down the activity of certain types of neurons, though there are still open questions about what regions are especially vulnerable to this inhibition. A different avenue of research studies the loss of awareness during deep sleep. Others study the absence of awareness of one stimulus when attention is paid to another. Difficult though it is, research on the interdependence of conscious and nonconscious processes is accelerating.

NAY-SAYING IS EASIER THAN DOING SCIENCE

IMPRESSED BY our ignorance, some philosophers have expressed certainty that *no* answer will be forthcoming—we will never know how the brain gives rise to thoughts and feelings. One popular reason offered for this is that no one can imagine what a detailed and satisfying neurobiological explanation would actually look like. So, the argument goes, the inability even to imagine an explanation is a sure sign that this is no mere problem, but an unsolvable mystery.[21] Those who nay-say in this vein are not necessarily dualists, though they are apt to share almost everything except the name with Cartesian dualism. To face up to this nay-saying, we pause now for a philosophical interlude wherein we deal with this roadblock: *Consciousness is too deep a mystery for us ever to understand. Give up trying.*

Two things can be said at the outset about the nay-saying. First, the argument embodies a very strong prediction: *no one will ever solve the mystery—not ever, no matter how science might develop.* Never is quite long. Longer than a lifetime. This prediction, tendered as obvious, is really spectacularly rash. After all, the history of science is chock-full of phenomena deemed too mysterious ever to be understood by mere mortals, but which eventually did yield to explanation. This could be just one more of those.

The nature of light was one such problem. Well into the nineteenth century, the scientific consensus was that light is a fundamental feature of the universe, never to be explained by anything more fundamental. What happened? By the end of the nineteenth century, James Clerk Maxwell had explained light as a form of electromagnetic radiation—on the same spectrum as X-rays, radio waves, ultraviolet waves, and infrared waves. So the prediction, seemingly certain and unassailable, was flat-out wrong. Interestingly, it is now difficult to find anyone who even knows about the confident but misguided predictions concerning the inexplicability of light.

It hardly needs noting that it is preposterous to infer that something is *unknowable* simply because it is *not known*—especially when the science is in its very early stages. So much about how nervous systems function is not yet fully understood, such as how memories are retrieved or attention allocated or why we dream. Imagine a prediction made in year 2 CE saying that no one will ever understand the nature of fire. Certainly, at that time no one knew the slightest thing about what fire really is. No one knew that there was such a thing as oxygen, let alone that fire was rapid oxidation. It was widely thought that fire was a fundamental element, along with earth, air, and water, not to be explained beyond describing its behavior. The problem was eventually solved, though not until about 1777 by the French scientist Antoine-Laurent Lavoisier.

Or imagine a prediction in year 1300 that science will never understand how a fertilized egg can end up as a baby animal. Or a prediction in 1800 that no one will ever succeed in making something that can control infections. Suppose someone predicted in 1970 that science could never find a way to record levels of activity in a normal human brain without opening the skull. Wrong. This technical achievement flowered in the 1990s as functional magnetic resonance imaging (fMRI) was developed. As a student of the brain in the 1970s, I would have been inclined to scoff at that possibility as science fiction because I could not *imagine* a device that would do the trick. My scoffing would have been merely an expression of my ignorance. So my imagination was not up to the job.

To the degree that the nay-saying rests on an unsubstantiated prediction, it need not deter us from moving forward.

Here is the second thing about nay-saying. There is something smugly arrogant about thinking, "If I, with my great and wondrous brain, cannot imagine a solution to explain a phenomenon, then obviously the phenomenon cannot be explained *at all*." Yet some philosophers and scientists find themselves strongly attracted to this assumption.[22] They should not be. Why would you take my inability to imagine a future development in science as a reliable index of whether and how a problem can be solved? After all, I may have a pinched imagination, or my imagination may be limited by my ignorance (there it is, *again*) of what science will unearth a decade or two hence.[23] What I can and cannot imagine is a psychological fact about me. It is not a deep metaphysical fact about the nature of the universe.

The flaws in the presumption that we will never explain mental phenomena in terms of the brain can also be analyzed from a slightly different perspective. An inference is supposed to take you from something for which you have very good evidence (such as an observation) to something else that is probably true. For example, I might infer that there is a forest fire on the other side of the mountain. My evidence is that I see firefighting helicopters carrying water to the other side of the mountain. Seeing the helicopter allows me to get new knowledge because

I can infer something new. How does the naysayer's inference concerning neuroscience look?

The naysayer's inference takes us from ignorance—we are *ignorant* of the mechanisms for conscious awareness—to knowledge—we *know* that conscious awareness cannot be explained. This spells trouble. Suppose your doctor says, "We are ignorant of what causes your rash, so we know it is caused by a witch." You would get a new doctor. Fast. Inferring knowledge from ignorance is a fallacy, and it is why the ancient Greeks labeled this fallacy an argument from ignorance. Here is another obviously fallacious version of such an argument: I do not know how to explain how monarch butterflies navigate to Mexico, so I know it is by magic. Nuts. Ignorance is just ignorance. It is not special knowledge of magical causes. It is not special knowledge of what can or cannot be discovered in the long haul of time.

It is conceivable that science will never understand how neurons produce feelings and thoughts. Nevertheless, you cannot tell just by looking at a problem that it is not solvable by science. You cannot even tell whether the problem is really hard or downright tractable. Problems do not come with levels of difficulty pinned to their shirts. Moreover, as science progresses, the shape of a puzzle often begins to change, and some scientist somewhere may look at the problem in a new way, or some unforeseen technological development may render the problem quite tractable. Here is an example to illustrate the point. In the early 1950s, many scientists thought that solving the problem of how information is inherited by offspring from parents—the copying problem—was really, *really* hard, perhaps unsolvable. On the other hand, the problem of explaining how a protein molecule takes its typical three-dimensional shape once it is made was thought to be relatively easy. It turned out to be exactly the other way around.

In 1954, James Watson and Francis Crick published their paper explaining that DNA is a double helix, with orderly sequences of pairs of bases that look a lot like a code. This monumental structural discovery was the key to the solution of the copying problem, the details being worked out over the next decades. By 1975, every biology textbook explained the basics of genes, how DNA codes for proteins, and how proteins get made. What about the allegedly "easier" problem of how proteins *fold* into their three-dimensional shape once made? That answer is still a research project.

This smackdown has been a little drawn out for the following reason: a number of philosophers, most famously David Chalmers, made their reputation by (a) claiming that the nature of consciousness cannot be solved by studying the brain, (b) giving the problem a name (the *hard* problem), or (c) claiming that consciousness is a fundamental feature of the universe, along with mass and charge.[24] No equipment had to be designed and maintained, no animals trained and observed, no steaming jungle or frozen tundra braved. The great advantage with naysaying is that it leaves lots of time for golf.

EXPANDING OUR SELF-CONCEPTION

SO PROBABLY the soul and the brain are one and the same; what we think of as the soul is the brain, and what we think of as the brain is the brain. Can we still talk about a great-*souled* person and not mean a great-*brained* person? Can we still say that a tennis partner put up a *spirited* defense or that the *spirit* of the times is coalitional, not confrontational? Sure, why not? We know what we mean, because these are conventional expressions. Similarly, we may still say that the sun is setting, even when we know full well that Earth is turning. Soul music, soul food, and having soul are still what they always were.

Can a person live a spiritual life even if there is no soul in the Cartesian sense? Or if you no longer believe you have a soul?

For simplicity, let's assume that *spiritual* means valuing certain kinds of things, such as reflection, passing quiet time in a quiet place, choosing to do some things slowly and thoughtfully, not hectically and anxiously. It may mean not valuing money and its power overmuch and finding contentment in simplicity. Notice that for none of these values and preferences do we need to rely on the idea of a nonphysical soul. Would certain lifestyles be better or more valuable if we *did* have a nonphysical soul? Not that I can see. How would having a nonphysical soul make enjoying silent periods better? Part of what is going on when you enjoy your spiritual times is that your brain detaches from certain worries and woes; your breathing slows, facial muscles relax, you let go.

Put this in the context of a yoga practice (or substitute running or chanting, if you prefer). I often experience semi-euphoric feelings after yoga class, during the meditation period at the end (savasana). Later, I wonder about what happens in the brain during practice such that the blissful feelings typically follow in savasana. Why does this lovely feeling reliably occur?

One hypothesis is that during yoga exercises, the intense focus on the task of getting the body into the right position causes a shift in the typical "oscillating balance of power" between two general systems of the brain: the *task-oriented* system (for example, when you ponder why your computer lost its Internet connection) and the so-called *default* system, a kind of "inner reflection" or "mind-wandering" system (for example, when you go over a conversation you just had or plan for the next one). This latter is engaged when you briefly switch off-task and worry about the next job or fret over the past one or fantasize about something (like sex, for example).

Using brain-imaging technology, researchers have shown that during meditation, the activity level of the self-reflection regions does decrease, and this occurs across various meditation routines.[25] This finding is consistent with the claim that meditative practices can increase feelings of peacefulness, contentment, and joy. I am guessing that many other practices that involve focusing attention away from worries and on the present task—prayer, chanting, running, playing golf, playing in a quartet—have similar effects.

Although there is a suggestive correlation between deactivation of the default network and feeling peaceful, this is not yet a demonstration of causality. Moreover, devoting a greater proportion of attention to on-task versus mind wandering may be only a small part of the story of postpractice bliss. For one thing, the value of spending more attention on the task likely depends on the task. If the task is one you do unwillingly, such as cleaning out horse stalls, then spending more time in the self-reflection (default) network having sexual fantasies is probably more conducive to joy.

So do I think of the joyful experiences during or following these practices as spiritual? Yes, I do, because they resemble experiences that other people describe as *spiritual*, including people who, like me, do not think we have a nonphysical spirit that we call forth during these practices. That the experience involves particular aspects of brain circuitry and chemistry is neither here nor there so far as the quality of the experience itself is concerned. Consider that people who take lysergic acid (LSD) or mescaline are apt to suppose that their remarkable experiences involve visiting a completely different but real world, a spirit world. Conviction of practitioners notwithstanding, the LSD and mescaline experiences are clearly brain-based phenomena. In both instances, the drug molecules bind to specialized serotonin receptors in the brain, changing the response patterns of the neurons. Understanding a remarkable experience, whether occasioned by drugs or meditation, within the framework of neuroscience makes no difference to the quality of the experience as such, but only to how we make sense of it.

THE CONCEPT of the soul, though having a long and respectable history, now looks outmuscled and outsmarted by neuroscience, so far as explaining our mind and our behavior. Yet one important motivation for favoring the concept of the soul has stemmed from the possibility of an afterlife. Moreover, reports of life after death and out-of-body experiences abound. Surely, one might think, these accounts put the soul back in the story. Possibly. Let's first find out how plausible these accounts are and what they really tell us.

Chapter 3

My Heavens

HEAVEN IS FOR REAL, or so it has been proclaimed. Here is the kind of evidence offered in support of heaven: Alex Malarkey (actual name) died—*really* died—and then came back to life and reported what he observed in heaven—angels and Jesus. His father, Kevin Malarkey, wrote the book *The Boy Who Came Back from Heaven*. It was on the best-seller list for many, many months. Afterlife is also an idea explored in the movie *Flatliners*, where medical interns decide to test the heaven hypothesis by "flatlining" themselves (when there is no heartbeat, the cardiac monitor shows a flat line) and then "coming back from death" to report what they experience. Of course, that was a *movie* and not an actual test.

To judge from the books available on "after-death" or "near-death" experiences, people have come back from the dead and reported lights, divine beings, dead relatives living in another realm, and so forth. Some people intensely and avidly want to believe in heaven. I intensely and avidly want to know what is probably true, or what real evidence shows—about my bank account and the state of my teeth, *and* about life after death. Not all alleged evidence is reliable evidence; some is wishful think-ing, some is hyped up, some is just a moneymaking scam. Let's look closer.

Following cardiac arrest, a patient may be resuscitated; the heart may beat again, and breathing may be restored. Later, a small minority of these patients may describe an experience involving peaceful feelings and perhaps visual perceptions of tunnels, lights, and even angels. This has become known as the near-death experience. A popular interpretation of the reported experience goes like this: The patient was actually dead for a brief spell, the soul began the transition from the body to heaven (or the next world), and the patient saw his or her dead loved ones at the end of a bright tunnel. Then the patient was brought back to life on Earth. The experience was transformative, and thereafter, the patient was no longer afraid to die.

Maybe heaven is for real, but maybe not. As with any hypoth-esis, I want to take a careful look. Why? Well, because truth and living with truth is very important to getting on in life. Not *truthiness*, but truth. As Suze Orman says in her lectures on money, "Stand in the truth. Do the arithmetic, and stand in the truth. If you cannot afford a vacation, tell that to your kids and stand by what you say."[1] In the very short run, really believing something because you simply *want* it to be true can be comforting. In the long run, it is generally catastrophic. Of course, we live with uncertainty every day, but some things are more certain than others.

When I was still too young to work a full day in the fruit-packing plant, I decided one summer to join my friend Sandy in "Bible school." This summer event was not associated with our church or any other church in the village, but was run by a kind of roving preacher who held the classes upstairs in the Elks Hall. Attending was something to do, since real school was out.

The preacher began the morning lesson by describing hell—vividly and menacingly. This turned out to be a distinctly nasty place, and we were asked to remember having had a burn and then to imagine ourselves completely covered in horrible burns. We imagined ourselves, wicked as we knew ourselves to be, burning in hell, year after year, for eternity. Apparently, there were devils hanging around, laughing at our ordeal. No amount of screaming would do us any good. The pain would never cease, and so on and on.

As it happened, I had firsthand knowledge of third-degree burns. When I was 3, I was playing around in the orchard, and as usual for the summer, I was in bare feet. I wandered, unknowingly, into an ash pile that my mother had dumped from the woodstove that morning. The ash pile was gray, showing no evidence of fire, but it was actually terribly hot. Confused and alarmed, I stood screaming in the ash pile until rescued by my sister, who rushed down from a ladder where she was picking peaches. My feet were horribly burned. By age 10 when I encountered preacher Joe, my feet were fine, but I remembered well what burns were like.

My father, when queried about the veracity of the damnation-for-the-wicked story, was forthright: "Rubbish," he said, "pure nonsense. He makes it up. Besides, why does he think you are wicked? Mischievous and stubborn perhaps, but not evil or wicked." My mother took the view that you can expect this sort of tripe from traveling evangelists, who liked to scare people out of their wits to win converts. "They should get a real job," she, the Christian, grumbled. The hellfire and brimstone picture did not fit at all with their down-to-earth approach of community and good works and common sense.

The next morning, I did not go back to the Elks Hall, but instead took our dog Ferguson down to the creek to look for crayfish under rocks. Not all religious people are to be trusted, I explained to Ferguson. And if the hell of the traveling preacher is rubbish, what about the heaven referred to by our own gentle if boring Reverend McCandless? And seriously, how could bodies go up somewhere when they were clearly rotting in the grave? Ferguson, always lovingly attentive, cocked his head to listen and went back to nosing around for gophers.

Here is what I want to know: Were all the people who reported a near-death experience *actually* dead or merely near death? One thing to keep in mind is that even once the heart has stopped, there may be residual brain activity for a short period—longer if oxygen is supplied. *Were the subjects actually brain dead?*

Brain death is taken to imply that the critical brain regions for sustaining heartbeat and breathing (regions in the brainstem) no longer function, and the patient shows no brainstem reflexes, such as the pupil contracting when a light is shone on the eye.[2] Some 25 different assessments are used to determine brain death. If, for example, there is some residual brainstem and cortical activity, then the brain is not dead. Then odd experiences may result.

In the Gospel of St. John, Lazarus is said to have been stone-cold dead for *four* days. John was not there to report, but he wrote of the story much, much later, as it was handed down by word of mouth. John says that Lazarus "stinketh," meaning that the body was already in decay. Yet, according to John, Jesus miraculously brought Lazarus to life again. So far as I can determine, there are no credible reports of someone who is actually brain dead, and certified appropriately by a physician as brain dead, who has actually woken up as Lazarus did. Consequently, what must be assumed in the near-death literature is that being *nearly* dead is enough to get a peek at the afterlife. That this is an assumption is not always made clear. Nor is it explained why near death is sufficient to allow that heavenly peek.

In the book written by Kevin Malarkey, there is no mention that Alex's brain was ever scanned or even monitored for activity with scalp electrodes to determine whether he was brain dead. There does not appear to be any reason to think that he was brain dead. It is obvious that he was in coma, which is not at all the same as being brain dead. Quite a lot of activity was likely going on in Alex's brain following his head trauma, especially during the period as he recovered and his coma became less profound. Likely the cortex became more active as he began to emerge from coma. Is this the period when he saw Jesus? We do not know, but it seems likely. His vision of Jesus during his emergence from coma was probably no different from a dream of Jesus or a fantasy of Jesus.

One fairly reliable way to assess the prospects of a patient in coma after severe anoxia, such as in drowning, is to image the brain twice, separated by several days. If the brain is severely damaged, substantial shrinkage will be observed in the images over the course of a few days. Shrinkage is seen more frequently in severe oxygen deprivation (anoxia) than in head trauma. When shrinkage is seen, the prospects for recovery of function are vanishingly small. In children, determining brain death requires two assessments separated by a time delay of several days, with the second assessment performed by a different physician than the one who performed the first assessment.

To the best of my knowledge, not a single patient who has been credibly diagnosed as brain dead according to the aforementioned criteria has come to consciousness and reported seeing dead relatives, divine persons, or angels while brain dead. This suggests a more modest interpretation of experiences of life after death: patients who report "coming back from the dead" were not in fact brain dead, though they must have suffered other effects that prevented them from emerging into full consciousness, such as lowered levels of oxygen and some swelling of the brain.

Anoxia tends to have a poorer prognosis for recovery than head trauma, though in cases where the person was very cold— for example, by being in icy water—the low temperature can be somewhat protective of brain cells. Children who suffer head trauma may be in coma for some months and show significant recovery of function. This was the pattern in Alex Malarkey's case. He was in a car accident and suffered a severe head injury. He was in coma for 3 months. Children in coma for 12 months have a poorer prognosis for recovery than those who emerge from coma after 3 months. Is coma brain death? No. While a highly undesirable condition, coma is not brain death.

Criteria for coma include the lack of responsiveness to external stimuli, including a pinprick, lack of bowel and bladder control, and no sleep/wake patterns.[3] After the initial crisis, patients in coma show brainstem reflexes, such as the pupils contracting to a bright light, unlike brain-dead (irreversible coma) patients.

Just to round out the complexity of the range of conditions, I should add that patients who are diagnosed as being in a vegetative state also lack responsivity to external stimuli and do not follow a light even if their eyes are open. They also lack bowel and bladder control. But they do show sleep/wake cycles. This was the condition of the tragic patient Terri Schiavo. At autopsy, her brain was found to be severely damaged, especially in cortical regions. She would never have recovered consciousness even if she had been kept on life support machines indefinitely.[4]

Minimally conscious (MC) patients, by contrast, tend to be able to track a light with their eyes and may show responses to a mild pinprick on the hand or leg. The prognosis for recovery is better for an MC patient than those in a vegetative state. But biology is biology, and there are significant individual differences, as well as effects of age, overall health, and conditions such as diabetes and heart disease.

Brain tissue is highly sensitive to oxygen supply, and even very short periods of lower oxygen levels (hypoxia) can trigger unusual feelings and perceptions. People with low blood pressure who suddenly stand up from a squat are apt to experience a ring of blackness enlarging around a decreasing circle of light—like a tunnel. I often experience this myself and will feel a bit woozy for a few seconds until the blackness recedes and the tunnel expands to become my normal vision. Might the tunnel and light described by someone with low blood pressure be not unlike the experience of someone suffering anoxia during cardiac arrest, for example? Should I have thought I was experiencing heaven and strained to see angels or Great-Grandad Couts?

During cardiac arrest, the oxygen supply to the brain is diminished. Nerve cells are very vulnerable and begin to die when deprived of oxygen, even for a minute or two. Additionally, being in mortal peril and close to death may also trigger the release of endogenous opioids, something that is known to happen on the battlefield or during times of high danger. This can produce feelings of peacefulness and even elation. This release of endogenous opioids can also happen when a female is giving birth, with feelings of elation persisting well after the delivery.

If someone is near death, the main task is to resuscitate the patient, not to conduct research. Nevertheless, some studies have arranged for staff to document how many subjects with cardiac arrest have unusual experiences, how similar these experiences are across patients and across cultures, and the degree to which cultural beliefs affect interpretation.[5] For example, since Alex was growing up in a devoutly Christian household, it is hardly surprising that the divine person Alex believed he encountered was Jesus, not Confucius or Allah or Zeus. Buddhists do not have visions of the Virgin Mary, Mormons do not have visions of Venus—or if they do, these are not considered religious experiences but hallucinations.

FUNNY THINGS CAN HAPPEN IN THE BRAIN

RELEASE OF endogenous opioids made by your brain can evoke positive feelings, including perhaps feelings of peacefulness.[6] Endocannabinoids, cannabis-like molecules recently identified in the brain, can also reduce anxiety and increase pleasant feelings. Such feelings, along with a sense of deep connectedness to everything, can also be brought on by LSD, psilocybin, and mescaline, as well as of course by opium and cannabis.[7] Incidentally, alcohol also produces its pleasant effects by stimulating the release of endogenous opioids.

Out-of-body and other dissociative episodes can be induced by ketamine, the drug sometimes used in anesthesia. According to my students, ketamine has now become a party drug, allowing them to explore the sensation of floating above their body, having their mind vacate their body, and so forth. That such purely physical interventions can induce experiences qualitatively similar to those called *near-death experiences* speaks strongly in favor of a neurobiological basis, one that can eventually be discovered through research.

Next question: How similar are the experiences across subjects? Variability is standard, and this suggests that there is variation in the patients' conditions and in their brains. In one large study of 344 patients who suffered cardiac arrest,[8] only about 18 percent (62) reported a core near-death experience (tunnel with light, felt dead, felt peacefulness). Others may have had an experience but no memory of it, because, of course, the study relied on post-resuscitation reports. Only 50 percent of the 62 patients who did have near-death experiences had the feeling that they were dead; only 56 percent (35) of the 62 patients had positive feeling such as peacefulness; only 24 percent (15) had the experience of leaving the body; only 31 percent (19) had the experience of moving through a tunnel; only 23 percent (14) had an experience with light.

If heaven really awaits us all after death, it is a little puzzling that only 12 percent of resuscitated patients reported a near-death experience with tunnels and light and peacefulness.

Is it plausible that there will be a neurobiological explanation of the cluster of phenomena? As neuroscientist Dean Mobbs and his colleagues pointed out, a strong reason for saying yes is that experiences similar to those suffering anoxia following cardiac arrest can be induced by electrical stimulation of the temporal lobe and hippocampus, something that may be done in epileptic patients prior to surgery. Similar experiences can also be induced by raising the level of carbon dioxide (hypercapnia, sometimes suffered by scuba divers) or by decreasing oxygen levels in the brain by hyperventilating following the Valsalva maneuver (as when you strain at stool).

To be sure, in contrast to hyperventilating following Valsalva, nearly dying may be transformative, for then the end of life cannot be very far away, and the reality of death is starkly evident. In such circumstances, people will be inclined to reevaluate their lives, relive memories, and reexamine what matters in life. They do this regardless of whether they believe in heaven or not. That very reexamination may be the transformative element in those who do have an unusual experience. The experience merely triggers the reflection that is transformative.

After a vigorous yoga practice when I am lying with eyes closed in relaxation pose (savasana), I sometimes feel that my body is floating just a few inches above the floor. It is a pleasant feeling of being very light. The early part of the relaxation procedure consists in focusing on each body part in turn and thinking about that body part relaxing, letting go. Once that is done, I feel as though I am floating just a bit off my mat. Were I of a different mind-set, I might imagine that I *really* am floating, perhaps in some spiritual domain. I might imagine that my spirit had defied the law of gravity and up floats my body. As it is, I enjoy the feeling, which may in fact be a kind of vestibular hallucination plus a little endogenous opioid to yield the pleasant sensations. The straightforward test for whether I ever actually float is this: Has anyone in my class ever seen my body float while I am in savasana and I feel as though I am floating? No one.

I think that some yoga practitioners would call this feeling spiritual, and as I indicated in Chapter 2, I am fine with that. I like the idea of a brain-based floaty feeling. So long as we do not mean that there is spooky stuff that makes my body literally lift off the floor.

HALLUCINATING BUT NOT DELUSIONAL

BRAINS DO funny things, like hallucinate. Everyone hallucinates.[9] You hallucinate every night when you dream. You are naked, trying to cover yourself, desperately seeking your lost luggage and waiting for the bus, which seems to be coming down a different street and will not stop for you anyhow. Everything keeps going wrong. In truth, as you know when you wake up, nothing like this is really happening.

A hallucination is a sensory experience of objects or events in the world that seem entirely real, but are not. The events in your dream life are merely the results of brain activity. When you awake, you are generally clear that the dreams were totally unreal.

Your dreams are often emotionally powerful, and this fosters the idea that they have special meaning regarding your real nature or the future or perhaps the "spirit world." Something more mundane seems to be true. Dreams are just emotionally powerful hallucinations, sometimes linked to events or feelings that occurred in the recent past, but full of random stuff, including dead relatives, flying horses, and talking mice, interestingly enough. The emotions, such as fear or anxiety, may be the more basic phenomena in dreams, and the random visual perceptions attach themselves to the feelings. Sometimes dreams provoke us to reflect on events in our real life, but dreams as such are not quite as meaningful as prophets and spiritualists of the dualist persuasion wish them to be. The reflections they provoke during the awake state, however, may be very meaningful and have a powerful effect on life decisions.

All mammals dream, so far as we can tell, and perhaps all vertebrates. I see my dog Duff, mutedly barking, legs twitching, mouth in a partial grimace. H e is hallucinating. Why all mammals dream remains a puzzle. Some neuroscientists think it is part of housekeeping, the generation of random patterns of activity, where only the patterns that link with daytime memories are tagged for storage. The evidence that sleep and dreaming are linked to learning and memory is becoming stronger, but exactly how is still puzzling (see also Chapter 9).[10]

Hallucinations in the awake state are rather more unusual and usually signify a medical problem. A host of different conditions may be involved, including drugs, tumors, seizures, migraines, sensory deprivation, and psychiatric diseases such as schizophrenia and dementias.

Just to round out the picture, here are some unusual, but not totally rare, examples of neural funny business. Kenneth is a 70-year-old successful furniture maker who explained sheepishly to his doctor that at many times during the day, he would have a very robust sensory experience of an extra arm and leg on his left side. His extra (supernumerary) limbs seemed to come and then, after a while, go. During the time when the extra limbs felt robustly present, they moved concordantly with his actual left limbs; third leg walking along, third arm swinging; left arm reaching for a biscuit, third arm reaching for a biscuit. He said that he could not direct his supernumerary limbs to move apart from moving his actual left limbs, and the presence of the supernumerary limbs did not trouble him too much, though of course it puzzled him. What on earth was going on?

Brain-imaging technology indicated an unusual form of epilepsy with a focus in the right temporoparietal junction (Figure 3.1). This result was helpful clinically because Kenneth was given medication that effectively suppressed his third arm and leg. But why would the supernumerary limbs appear at all? Another brain puzzle.

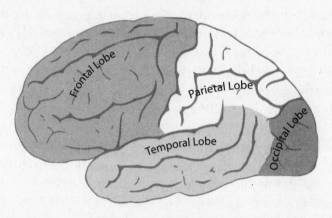

3.1 *The major divisions of the cortex. Note the junction between the parietal and temporal lobes.* Adapted from *Gray's Anatomy*, public domain.

Other variations in body sense may be correlated with unusual brain activities linked to migraine. When we were in high school, my friend Eunice suffered a weeklong catastrophic migraine every spring. In the days just prior to the onset of the migraine, she said that at night she usually had the experience of feeling that she was very, *very* tiny—about the size of a hornet—in a massively huge bed. The sensation did not alarm her because she knew it was a "brain thing" that would disappear. And it did. What did bother her was that her miniaturization experience reliably augured the migraine to come.

Other migraine "auras" involve visual effects such as a jagged cutting up of visual perception, sometimes just in one region of the visual field, such as upper right. Sometimes subjects see scintillation in a section of the visual field. On the first occasion of such an "aura," you tend to wonder, "Am I having a stroke?" Or you might wonder whether you are making a supernatural contact. Neither. Just a common, well-known, pre-migraine effect. No aliens or spirits gone berserk here.

Blaise Pascal, the French mathematician and philosopher (1623–1662), described these visual effects, as well as blindness in half his visual field prior to the onset of migraine headaches.[11] Eventually, he had an episode where he experienced a great deal of light in his visual field. He interpreted the light as a sacred visit and became a religious convert. Sometimes an aura can involve an unusual smell, such as scorched corn. Auditory hallucinations are exceedingly rare in migraine auras, but they also can occur.

Lila became blind as a result of age-related macular degeneration, and to her surprise, she began to experience visual hallucinations, mostly tiny people, sometimes cartoon characters like Elmer Fudd or Betty Boop. (The macula is a region of the retina densely packed with light-sensitive cells and is essential for clear and detailed vision.) Lila knew full well that the tiny people were not real, but she wanted an explanation for these experiences.

Lila's condition is known as *Charles Bonnet syndrome*, and it occurs in about 13 percent of patients with macular degeneration. (The physician Charles Bonnet reported this condition in the eighteenth century.) Like Lila, subjects with this condition are not delusional. They understand that these Lilliputian figures are not really there. The experience may be annoying nonetheless. It is suspected that Charles Bonnet syndrome may be underreported, as patients are apt to withhold such information, worrying that the doctor will categorize them as delusional. Rarely, a patient temporarily using an eye patch may also experience hallucinations, but this experience ceases with removal of the eye patch. (Eager for the experience, I have tried the eye patch myself, but, alas, without success.)

The explanation for Charles Bonnet syndrome is believed to involve sensory deprivation, where the brain, deprived of its normal sensory input, is activated by random activity and makes coherent, if odd, perceptions out of that activity. Such a vague explanation does not really explain much, and the syndrome remains poorly understood at the neurobiological level.

Remember when you tried to scream or run in a dream? Alarmingly, barely any sound comes, and your legs are leaden. Why can you not run, literally run, away from the fire-breathing monster? There is a brain answer: because there is a special bundle of neurons in the brainstem that makes sure that you cannot move during dreaming. In the dream state, these brainstem neurons actively block any motor signals that might emerge from the motor cortex, destined for the spinal cord and then the arms and legs. In effect, during dreaming you have a kind of temporary paralysis that actually prevents you from acting out your dreams. This was first discovered in cats by a French neuroscientist, Michel Jouvet, who showed that if the special brainstem network is damaged, then the cat will indeed jump up during dreaming and will run around as though chasing something.

This special brainstem inhibitor explains why you cannot scream but only croak pitifully when fending off attack during your dream or why you cannot kick your dream tormentor or run after your dream taxi. It is why your dog, when he is dreaming, makes funny muffled barking sounds and not the full-throated normal bark. The inhibition of motor signals may not be complete, however. Sometimes small twitches, chuckles, and whimpers are still possible, but normal movements are not. If, however, that inhibitory network is damaged, then like Jouvet's lesioned cats, you will act out your dreams. This sometimes happens in elderly subjects with dementia. Such a condition is extremely dangerous, since dream enactment is likely to result in injury to oneself or others. One subject, for example, leaped out of bed and ran at full tilt into a wall, injuring his head and damaging the wall. Another subject strangled his bed partner. In such subjects, REM suppression drugs are used to prevent the dream states altogether and thereby reduce the probability of injury.

In a small percentage of people, the usual synchrony of turning *off* dream paralysis and turning *on* wakefulness can be a bit sloppy. This means that paralysis may continue briefly after you are fully awake. Should this happen, you will likely be dismayed and frightened, fearing a stroke or capture by an alien. You are awake, you intend to move, but you cannot.

In general, these events are rare in a person's lifetime, but some people may experience them frequently, resulting in a fear of going to sleep, which in turn is associated with even more episodes of sleep paralysis intruding into waking life—a vicious cycle. Once sleep is deprived, subjects may also have frightening experiences, such as feeling the presence of an evil thing, hearing heavy footsteps clumping in the room, hearing menacing voices talk nearby, and so forth. This can also be induced in healthy subjects by regularly interrupting their sleep so they are sleep deprived. Not surprisingly, these exceptional events are frequently interpreted as encounters with the spirit world, ghosts and witches, or the dead. In subjects who come to the clinic for help, the intervention turns out to be remarkably simple. Get more sleep. It works.[12]

This list of perceptual oddities reminds us that the brain may do surprising things, things that have no special significance regarding afterlife or past life or spiritual life. They are just *neuro-oddities* for which we do not—not yet, anyhow—have complete explanations. Fascinating, poorly explained, sometimes annoying or disturbing, they are what they are: oddities.[13]

TAKING IT ON FAITH

NONE OF this discussion constitutes a knockdown demonstration that heaven does not exist. I cannot say that heaven does not exist with quite the same the certainty that I can say that leprechauns do not exist. Nevertheless, I have been unable to find in reports of near-death or so-called *after-death* experiences any credible evidence for such a place. The data on unusual experiences produced by unusual brain events prompt skepticism about claims that unusual experiences are genuinely of unusual things, such as ghosts and aliens and heaven. If, regardless, you want to believe in life after death, such a belief may be held as a conscious decision. You might say, "I want to believe in heaven because I feel better if I do." That might be fine. For me, I worry that such a decision might make me more vulnerable to flimflam. Self-deception can be like a drug, numbing you from much-needed feeling, weakening you exactly when you need to muster your resources for getting through a difficult reality.

Is some self-deception good for us? Possibly, but self-deception can run amok. A farm friend of my parents, charming and delightful Stanley O'Connor, was diagnosed at 58 with stomach cancer. In 1955, nothing could be done except to make him comfortable as the end approached. But Stanley did not give up. He and his wife withdrew their entire savings from the bank and borrowed an additional sum to boot. They packed up their old truck and drove south across the U.S. border to seek a cure from a semi-famous traveling faith healer who warmly received their "offering," prayed enthusiastically, laid on his hands, and then sent them packing. Three months later, Stanley was dead and the money was gone.

When I asked my father how Mrs. O'Connor now felt about the faith healer, he shook his head sorrowfully and explained, "Well, she believes her faith was not strong enough. Had it been, Stanley would have been cured. That is what she says." Around our kitchen table, this was discussed long and often, and it was seen as self-destructive self-deception. The faith healer, my father had always argued, was a charlatan, making money by preying on the desperate and ignorant. "Worse than dying," he opined, "was to go down stupidly." With no money to run the farm, support the children, or pay down the debt, Mrs. O'Connor herself fell into ruin and died the following year. Three young children, bereft and confused, were shipped out to reluctant aunts and uncles.

Several years later, in 1960, the movie about the traveling tent evangelist Elmer Gantry was released.[14] I found it almost unbearable to watch. Poor, ignorant farmers, desperate and full of hope, were easy marks for a charismatic huckster.

Stand in the truth. So counsels Suze Orman, and it is a powerful, lifelong message. It echoes the words of Aristotle, Confucius, Ben Franklin, and Mark Twain. And then there is Bertrand Russell, who said:

> Men fear thought as they fear nothing else on earth—more than ruin—more even than death. . . . Thought is subversive and revolutionary, destructive and terrible, thought is merciless to privilege, established institutions, and comfortable habit. Thought looks into the pit of hell and is not afraid. Thought is great and swift and free, the light of the world, and the chief glory of man.[15]

The facts do not modify themselves to conform to our beliefs, our hopes, or our dogma. Earth does not make itself the center of the universe because such an idea is deeply appealing to some humans; the heart really is just a meat pump. My children are my children because of the way reproductive biology works, not because an occult force preselected just *this* egg and just *that* sperm to get together and produce just *this* child. You can love your children fully without importing that bit of silliness.

The idea of a Platonic soul that departs the body for heaven when we die may draw us strongly to itself, but we have to ask whether that is true, whether that is truly what we believe, and whether we are merely fond of the idea without attaching significant credibility to it. How you think about this does make a difference to your decisions here and now—as it did with Stanley O'Connor. Will breastfeeding change a man from being gay to being straight?[16] Are the oceans being fished out?[17] Is climate change a hoax? It is tempting to have faith that God will provide. But it is unwise actually to believe it or to make that idea a reason for turning a blind eye to the tough matter of gathering evidence and thinking. God often fails utterly to provide.

Philosophers Tim Lane and Owen Flanagan have suggested that for a very small segment of our lives, mainly having to do with a certain subset of diseases we may have, false optimism may actually be beneficial: diseases such as cancer, but not diabetes or ulcers.[18] Believing that you will recover might—conceivably—help reduce the level of stress hormones, giving the immune system and modern medicine a better chance to do their work. What is the evidence? Decidedly iffy. Grouchy cancer patients seem to recover at about the same rate as cheerful patients, and there are no studies where some cancer patients rely only on positive thinking while the others get the standard medical treatment.[19] Nor is any such study likely to be funded.

Lane and Flanagan wisely point out, however, that false optimism is generally less a belief than a kind of belief-desire amalgam, more akin to hope than to a firmly held, evidence-based belief. It is something a wise person recognizes as unsupported by the cold, hard facts, but something that might end up being modestly palliative for this very specific condition. But the operative word here is *might*, since the evidence that false optimism is good for you is, at this stage, largely wishful thinking. In any case, it is useful to remember that false optimism can end up being disastrous if it sends you to the faith healer instead of the cancer clinic, or worse, if it means that your child is not vaccinated against measles and polio because you think your love provides a protective shield.

———————

A DEEP respect for the truth, however hard that truth may seem at times, is something lauded in many cultures—among the ancient Greek philosophers such as Socrates and the ancient Chinese philosophers such as Mencius; among the Inuit, the Cheyenne, and the Trobriand Islanders. Among the hard-bitten farmers in the Okanagan. More often than not, blind faith turns out to be more dangerous than whatever truth is out there.

Chapter 4

The Brains Behind Morality

THAT WINTER IN NORTHERN Manitoba was harsh, and even small game had long been scarce. In the Chipewyan camps, the children were becoming weak with hunger. The elders remembered such years and feared deaths. Early one morning, a man left with his young wife and infant, snowshoes marking their long trek over the birch-lined hill. As winter wore on, there were deaths, and still no game to bring to the camp. Slowly daylight hours lengthened. One evening, the young wife and baby returned alone, healthy and strong—healthier and stronger than those she had left behind in the camp. But she carried a look of wariness, and the wary eyes did not hide what had fed the two in the winter months away from the camp. Spring did come, and with spring came deer—lean, but enough to feed the camp and with good marrow to provide fat. The women kept their distance from the young wife and baby, and she slowly became listless and quiet. They understood what must happen before summer. To live on the flesh of another was to have a tainted spirit; to have tasted human flesh was forever to be weakened to human flesh again. They knew it as "going Windigo." The camp watched and waited as the river ice melted, while she became more detached. So it came to pass, on a night when the northern lights were billowing green and yellow and the wind was calm, that she curled silently down upon the ground, while the elder gently wrapped the bear hide about her head. He then quietly lay upon her head until she was still. The baby was stilled a few moments later.[1]

Living as we do in prosperity and plenty, this story sears the imagination. What we can imagine well enough is the overwhelming love for the infant and the unendurable pain of seeing the baby slowly starve as the milk dried up. Mammalian parents do make extraordinary sacrifices for offspring. You are a mammal, and your mammalian brain is organized with offspring a primary focus of care and concern. But how much would you sacrifice to keep your baby from starving? Of the prohibition, embodied as a cultural practice, against eating human flesh, I obsessively wonder, "So, what exactly was behind the elder's decision to kill the mother?" I notice that I do not wonder whether it was wrong for him to have done so, although I know full well it would violate a law of our culture.

The group would have been small, probably 20 to 30 people. The elder's wisdom, in the story, reached back into generations of lives lived and stories retold, back to the understanding that eating another human to stay alive is apt to distort one's self-conception and inhibitions and to alter all social relations in the group, making for dangerous instabilities and mayhem. Group survival in severe conditions means a low tolerance for risky behavior. The prohibition does not seem to me to be foolish or ungrounded. I find I cannot say whether the prohibition was actually wrong.

Why did the *baby* have to die? Depending as they did on uncertain game, the indigenous people of the far north faced every winter as a struggle to keep alive. In bad times, the thought of staving off starvation this way would not have been rare. Like all cultures, the Chipewyan conveyed the core of practical, hard-won social insight by means of a compact myth—in this case, going Windigo. *A Windigo infant, motherless, grows up tended by well-meaning kin; unavoidably, he is seen, always, through the murk of his uncommon survival; socially, always in a hole, with nowhere to go where his story does not follow.* And thus, when the winters were savage, the people of the North generally held their weak ones as they died, rather than give in to desperation survival.[2]

WHERE DO VALUES COME FROM?

VALUES ARE not objects like stomachs and legs; they are not in the world in the way that seasons or the tides are in the world. But nor are they otherworldly; rather, they are social-worldly, and we live and die by them. They reflect how we feel and think about certain kinds of social behavior. Morality is not in the world in the way that the seasons are in the world. But morality is certainly in your social world; it emerges from the positive feelings of humans toward courage or kindness and in the nega-tive feelings toward brutality or child neglect.

The values of self-survival and self-maintenance are not in the world either. But they are in the brain of every animal. And it is easy to see how the biological world came to be that way. Brains are built by genes. Unless the genes build a brain that is organized to avoid danger and seek food, water, and mates, the animal will not long survive nor likely reproduce. If you have no offspring, your genes do not get passed on. So if you had genes that built your brain so that you had no regard for your well-being, those genes would expire with the brain they built.

By contrast, an animal that is motivated to care about its own self-maintenance has a better shot at having offspring. And that animal's genes will spread in the following generations as its offspring survive and reproduce. So certain self-oriented values are favored by natural selection. Quite simply, animals who have genes that build brains that have self-oriented values do better than those with genes that build self-neglecting brains.

Moral values, however, involve self-sacrifice in the care of others. How did humans come to have moral values? How did it come to pass that humans care about their offspring, mates, kin, and friends? Self-care seems to be in conflict with other-care. At a deep level, it turns out that moral values, like self-caring values, are in your brain. How did such an evolutionary development come about? The basic answer is that you are a mammal, and mammals have powerful brain networks for extending care beyond self to others: first to offspring, then to mates, then to kin, friends, and even strangers. But how did those networks come to inhabit mammalian brains? How could they be favored by evolution?

THE MAMMALIAN LOVE STORY

THE EVOLUTION of the mammalian brain marks the beginning of social values of the kind we associate with morality. (This story is probably true of birds, too, but for simplicity I will leave birds aside for now, regrettably.) The evolution of the mammalian brain saw the emergence of a brand-new strategy for having babies: the young grow inside the warm, nourishing womb of the female. In reptiles, by contrast, the offspring gestate in eggs left in sand or tucked into a hole, for example. When mammalian offspring are born, they depend for survival on the mother. So the mammalian brain has to be organized to do something completely new: take care of others in much the way we take care of ourselves. So just as I keep myself warm, fed, and safe, I keep my babies warm, fed, and safe.

Bit by evolutionary bit, over some 70 million years, the self-care system was modified so that care was extended to babies.

Now, genes built brains that felt pain when the babies squealed in distress or when the babies fell out of the nest. And of course the babies felt pain when they were cold or separated or hungry, and then they squealed. These new mammalian brains felt pleasure when they were together with their babies, and the babies felt pleasure when they were cuddled up with their mother. They liked being together; they disliked being separated.

Fine, but why did mammals evolve this way of having babies? What was so advantageous about the way early mammal-like reptiles (*sauropsids*) made a living that set the stage for this whole new way of having babies? The answer probably has to do with energy sources—with food.

The first sauropsids that happened to be warm-blooded (*homeotherms*) had a terrific advantage: they could hunt at night when the cold-blooded competition was sluggish. Lizards and snakes depend on the sun to get warm and to be able to move quickly. When the temperature drops, they become very slow. Pre-mammals probably feasted on sluggish reptiles lying around waiting for the sun to come up, or at least they could forage without fear of reptilian predators. Sauropsids also managed well in colder climates, thus opening new feeding and breeding ranges.

Homeothermy requires a lot of energy, so warm-blooded animals have to eat a lot relative to fish and reptiles. The energy cost of being a mammal or a bird is at least ten times the cost of being a lizard.[3] If you have to take in a lot of calories to survive, it may help to have a brain that can adapt to new conditions by being smart and flexible. Biologically speaking, it is more efficient to build brains that can learn than to build genomes that build brains with reflexes for every contingency that might crop up during life. When a brain learns, wiring has to be added, which entails that the genome has to have genes that get expressed to make proteins to make the wiring to embody new information. Rigging up that cascade of events is much less complex than altering a genome so that it builds a brain that can know at birth how to react in many different circumstances.[4] Notice that using a learning strategy to tune up the brain for strategic survival also means that at birth, the offspring do not know much. Mammalian babies are dependent.

So learning is a great way to get smart and augment your chances for survival, and this requires circuitry that can respond to experience to a much greater extent than can the hardwired circuitry for reflexes such as the eyeblink reflex. Nevertheless, an expanded learning platform needs to work hand in hand with the old motivational and drive systems already in place, and it needs lots and lots of experience-dependent neuronal changes if it is to be smart. You want circuitry that will allow you to respond flexibly to events in the environment, remember specific kinds of events and generalize, plan and choose between options, and do all these things intelligently. Cortex—a special kind of neuronal architecture—turned out to provide the kind of power and flexibility needed for these intelligent functions. What is *cortex*?

The cortex, new with mammals, is a highly organized six-layered network of cells that lies over the ancient reflex-organized brain like a quilt over an engine (Figure 4.1). Lizards have a kind of networky mantle lying over their deep structures, too, but it is three-layered, and it shows none of the highly regular organization seen in the mammalian cortex. The ancient structures under the cortex continued to play a big role in mammalian behavior, but their domination of behavior began to weaken as the cortex expanded and increased its control over behavior.[5] The great expansion of the prefrontal cortex (PFC) is associated with greater flexibility in behavior and a greater capacity for self-control and problem solving (Figure 4.2). Rats are smarter than lizards, and monkeys are smarter than rats. By *smart*, I mean they can solve problems and are cognitively flexible.[6]

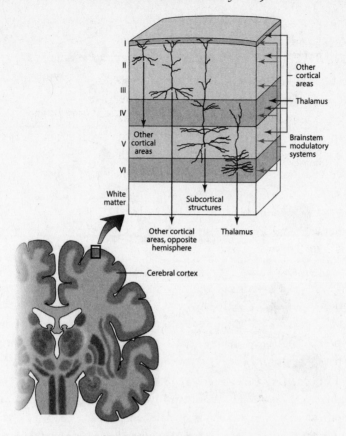

4.1 *The human brain as seen in coronal section (a cut from ear to ear). The gray edging on the outer surface is the cortex (cortical mantle). The difference between white matter and gray matter depends on the presence of myelin, which consists of fat-rich cells that wrap themselves around the axons of neurons, providing a kind of insulation resulting in faster signal transmission. Gray matter lacks myelin, consisting mostly of the cell bodies and dendrites of neurons. Other gray matter structures can be seen below the cortex. The cutaway depicts the cortex's laminar organization and highly regular architecture. Not conveyed is the density of the neurons: there are about 20,000 neurons in 1 cubic millimeter of cortical tissue, with about 1 billion synaptic connections between neurons and about 4 kilometers of neuronal wiring.*

Adapted from A. D. Craig, "Pain Mechanisms: Labeled Lines Versus Convergence in Central Processing," *Annual Review of Neuroscience* 26 (2003): 1–30. © Annual Reviews, Inc. With permission. Originally printed in Churchland, Patricia S. *Braintrust*. Princeton University Press. Reprinted by permission of Princeton University Press.

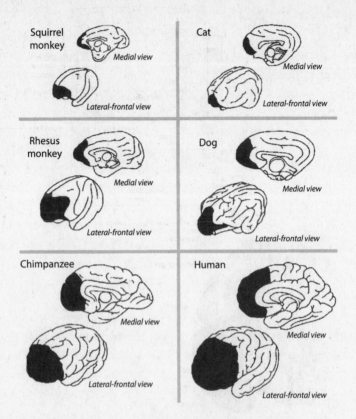

4.2 *The dark area corresponds to the prefrontal cortex in each of the six species shown. Two views are shown: lateral-frontal (as though looking from the side and front) and medial (so the extent of the prefrontal cortex on the inner aspect of the hemisphere is represented). Not to scale.* Reprinted from Joaquin Fuster, *The Prefrontal Cortex*, 4th ed. (Amsterdam: Academic Press/Elsevier, 2008). With permission from Elsevier.

Exactly how the six-layered cortex evolved is largely lost in our ancient past,[7] and unfortunately, no reptilelike mammals (sauropsids) survived. Nevertheless, comparisons of the brains of different existing species as well as studies of brain development from birth to maturity can tell us a lot. One thing we do know is that cortical fields supporting sensory functions vary in size, in complexity, and in the connectivity portfolio as a function of a particular mammal's lifestyle and ecological niche. For example, flying squirrels have a very large visual cortical field, whereas the platypus cortex has a tiny visual field but a large somatosensory (touch) field. The ghost bat, a nocturnal mammal that relies on precise echolocation to hunt, has a relatively huge auditory field, a small visual field, and a somatosensory field much smaller than that of the platypus (Figure 4.3).[8] Among rodents there are very different styles of moving—flying squirrels, swimming beavers, tree-climbing squirrels, for example. This means that there will also be organizational differences in the parts of the brain associated with skilled movement, including the motor cortex. In all mammals, the frontal cortex is concerned with motor function. In front of that is the prefrontal cortex—an area concerned with control, sociality, and decision making. All of these cortical fields have rich pathways to and from a whole range of subcortical regions.

Brains are energy hogs, and having a bigger brain means you need yet more calories just to keep your big brain in business. Moreover, helpless infant mammals eat an awful lot. This is because being immature at birth, their brains and bodies have a lot of growing to do. So pound for pound, mammals had to eat a lot more than reptiles. Being smart is an advantage if you need to catch lots of high-quality protein.

Being smart is also an advantage if you need to venture into new territory to find new resources or maybe migrate between territories as the seasons change. Because mammals eat so much more than reptiles, a given range supports fewer of them. Dozens of lizards can feed quite well on a small patch, but a patch that size will support fewer squirrels and even fewer bobcats.

Consequently, one impact of the big calorie consumption supporting homeothermy concerns reproductive strategy: mammals that bear just a few offspring and ensure that they flourish do better than those that breed like turtles, abandoning their many offspring to their fate.

Sensory domains

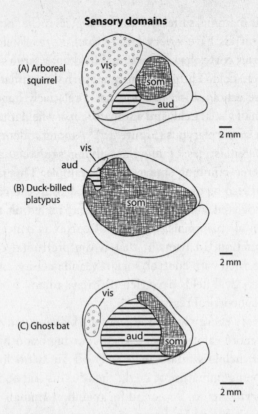

4.3 *The size of sensory fields in the neocortex of three different mammals with different sensory specializations. Sensory domains, or the amount of cortex devoted to processing input from a particular sensory system, are denoted in different shading. Note that* som *refers to the somatosensory cortex (touch, temperature, pressure),* vis *refers to the visual cortex, and* aud *refers to the auditory cortex. (A) The arboreal squirrel is a highly visual rodent, and much of its neocortex is devoted to the visual system. (B) The duck-billed platypus has an extremely well-developed bill containing densely packed mechanosensory and electrosensory receptors. The platypus uses its bill for most activities, including navigating in water, capturing prey, avoiding predators, and mating. Most of the neocortex in this mammal is devoted to the somatosensory system. (C) The ghost bat is an echolocating mammal that relies on its auditory system for most vital behaviors. It is not surprising that a large proportion of its neocortex is devoted to the auditory system.* Adapted from Leah Krubitzer and Jon Kaas, "The Evolution of the Neocortex in Mammals: How Is Phenotypic Diversity Generated?" *Current Opinion in Neurobiology* 15 (2005): 444–53. With permission from Elsevier.

So what do we end up with? Warm-blooded mammals that eat a lot, whose offspring are developed first in the mother's womb, then suckle at the mother's mammaries. We have helpless offspring whose brains are immature at birth but that grow much bigger as the offspring learn about their world, and who are nurtured by the mother and perhaps by the father to maturity. Nurturing raises a question: What in the brain motivates a mammalian mother to care about offspring if a turtle mother does not? Something has to be different. Mammalian mothers are attached to their infants; turtle mothers are not. Mammalian mothers take huge risks to themselves to care for their offspring; rattlesnake mothers do not.

If mammalian babies are helpless at birth, they have to be fed and defended. Why do mammalian mothers typically go to great lengths to feed and care for their babies? Two central characters in the explanation of mammalian other-care are the simple peptides *oxytocin* and *vasopressin*. The hypothalamus is a subcortical structure that regulates many basic life functions: hunger, thirst, and sexual behavior. (The hypothalamus is underneath the thalamus—hence "hypo"—and the thalamus is underneath the cortex. Anything in this neighborhood is said to be subcortical.) In mammals, the hypothalamus also secretes oxytocin, which kicks off a cascade of events so that the mother feels powerfully attached to her offspring. The hypothalamus also secretes vasopressin, which kicks off a different cascade of events so that the mother protects her offspring, defending them against predators, for example. Mothers love the babies and feel pain if they are threatened.[9]

If you feel attached to your babies, you strongly care about them and care for them. You want to be with them, and you feel uncomfortable when separated from them. Pain and pleasure, motives that worked well in reptiles and pre-mammals, are still potent drivers of behavior in mammals where they serve to generate new behaviors, such as parental care.[10]

The lineage of oxytocin and vasopressin goes back about 500

million years, long before mammals began to appear on Earth. What did these peptides do in reptiles? They played various roles in fluid regulation and in reproductive processes such as egg laying, sperm ejection, and spawning stimulation. In mammalian males, oxytocin is still secreted in the testes, and still aids sperm ejaculation; in females, it is secreted in the ovaries and plays a role in the release of eggs (ova). In mammals, the role of oxytocin and vasopressin in both the body and the brain was expanded and modified, along with wiring changes in the hypothalamus to implement postnatal parental behavior. These changes are part of the network organized to ensure that dependent offspring are cared for until they are capable of fending for themselves. This sort of modification of existing mechanisms for new purposes is a very familiar pattern in biological evolution.[11]

Simplified, here is how attachment and bonding works: Genes in the fetus and in the placenta make hormones that are released into the mother's blood (for example, progesterone, prolactin, and estrogen). This leads to a sequestering of oxytocin in neurons in the mother's hypothalamus (Figure 4.4). Just before the baby is born, progesterone levels drop sharply, the density of oxytocin receptors in the hypothalamus increases, and a flood of oxytocin is released from the hypothalamus.

The brain is not the only target of oxytocin, however. Oxytocin is also released into the mother's body during birth, facilitating the contractions that push the baby out. During breastfeeding, oxytocin is released in the brain of both mother and infant. Assuming the typical background neural circuitry and assuming the typical suite of other resident neurochemicals, oxytocin facilitates attachment of mother to baby—*and* of baby to mother. Gazing down at the infant on your breast, you feel overwhelming love and the desire to protect. Your infant gazes back, attachment deepening.[12]

Physical pain is a "protect myself" signal, and these signals lead to corrective behavior organized by self-preservation circuitry.

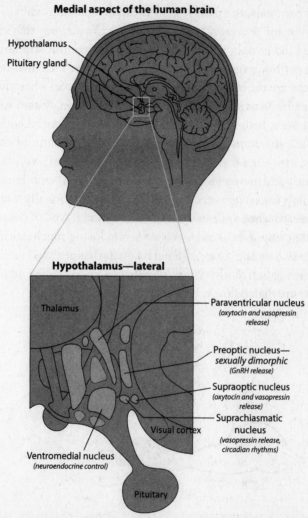

Medial aspect of the human brain

Hypothalamus
Pituitary gland

Hypothalamus—lateral

Thalamus

Paraventricular nucleus
(oxytocin and vasopressin release)

Preoptic nucleus—
sexually dimorphic
(GnRH release)

Supraoptic nucleus
(oxytocin and vasopressin release)

Suprachiasmatic nucleus
(vasopressin release, circadian rhythms)

Visual cortex

Ventromedial nucleus
(neuroendocrine control)

Pituitary

4.4 *The hypothalamus, though small relative to the cortex and thalamus, contains many regions (nuclei) that are essential for regulating self-maintenance and reproduction. It is also closely associated with the pituitary gland and the release of hormones. This diagram is highly simplified and shows only the main areas involved in sexual behavior and parenting behavior. Not labeled are areas associated with thermoregulation, panting, sweating, shivering, hunger, thirst, satiety, and blood pressure.* Modified from a Motifolio template.

In mammals, the pain system is expanded and modified; protect myself and *protect my babies*. In addition to a pathway that identifies the kind of pain and locates the site of a painful stimulus, there are pathways responsible for emotional pain. This involves a part of the cortex called the *cingulate* (Figure 4.5). So when the infant cries in distress, the mother's emotional pain system responds. She feels bad, and she takes corrective behavior. Another area called the *insula* monitors the physiological state of the body (the insula is a sophisticated *how-am-I-doing* area). When you are gently and lovingly stroked, this area sends out "emotionally safe" signals (*doing-very-well-now*). The same "emotionally safe" signal emerges when your baby is safe and content. And of course, your infant responds likewise to gentle and loving touches: *ahhhhh, all is well, I am safe, I am fed*. Safety signals dampen vigilance signals, which is partly why safety signals feel good. When anxiety and fear are dialed down, happiness is possible.

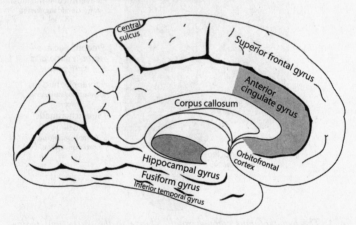

4.5 *A schematic diagram from the medial view of the human brain (as though one hemisphere were detached), identifying the location of the cingulate cortex, which wraps around the corpus callosum, the massive sheet of neurons that connects the two cerebral hemispheres. A gyrus is a hill; a sulcus is a valley. Shown also are some of the main subdivisions of the prefrontal cortex.* Adapted from *Gray's Anatomy*, public domain. Originally printed in Churchland, Patricia S. *Braintrust*. Princeton University Press. Reprinted by permission of Princeton University Press.

The expression of maternal behavior also depends on the endogenous opioids—those surprising substances normally synthesized by our own nervous system and whose actions can be mimicked by plant-derived drugs such as opium. This means that during suckling and other kinds of baby care, the opioids dampen your anxiety and make you feel good. Not high, but comfortable, safe, and satisfied. One way we know this is by experimentally blocking the receptors for the endogenous opioids. If you do that, maternal behavior is blocked, an effect observed, for example, in rats, sheep, and rhesus monkeys.[13]

A female fox with pups will be greatly distressed when her nest is approached or when a pup is removed and begins to squeal. Fear, anxiety, and arousal and an increase in stress hormones are now in play. Her vasopressin levels ensure that she will be vigorous in defense of her pups. When order is restored and the pup is back with its mother, endogenous opioids provide rewarding safety feelings. If a mammalian mother is successful in making the infant safe and content, endogenous opioids are released, both in the brain of the contented infant and in the brain of the relieved mother.

Although some mammals, such as marmosets and titi monkeys, are biparental (both parents help rear the young), in many species, such as elk, polar bears, grizzly bears, and foxes, the father takes no interest in parenting and shows none of the mother's attachment to the infant. As we shall see, moreover, there are many variations on the basic circuitry regulating parental behavior, depending on a species' ecological niche and how it makes its living. For example, sheep refuse to suckle any lamb that is not their own (as identified by smell), whereas pigs and dogs will often suckle non-kin and even infants of other species.

Studies on the effects of separating rodent pups from their mother (3 hours a day for the first 2 weeks of life) reveal experience-dependent changes in oxytocin and vasopressin synthesis, as well as changes in brain-specific regions of the oxytocin and vasopressin receptors. Behaviorally, the pups that were separated from their mothers showed heightened aggression and anxiety. In some way that is not yet entirely understood, the rats' brains and behavior were altered by being in a deprived social environment.[14]

Here is where we are in the values story: That anything has value *at all* and is motivating *at all* ultimately depends on the very ancient neural organization serving survival and well-being. With the evolution of mammals, the rudimentary self-caring organization was modified to extend the basic values of being alive and well to selected others—to *me and mine*. Depending on the evolutionary pressures to which a species has been subject, caring may extend to mates, kin, and friends. Social mammals, those that live in groups, do tend to show attachment and caring behavior to others besides their own offspring. They like to be together and feel pain if ostracized. Exactly which others come within the ambit of caring depends, as always, on the species and how the brain evolved in that species to enable the animal to make a living and pass on its genes.

To better defend themselves against lions, baboons live in groups. Baboon society is matrilineal, with mothers and their daughters forming ranking matrilines. Everyone in the troop knows the ranking of the matrilines and who ranks where within a given matriline. Attachment is strong within the matriline, weaker outside it. Chimpanzee society is very different, as young females leave the troop in search of a new home. The males are largely related to one another and are tightly bonded. In wolf packs, there is only one breeding pair, whereas among orcas and seals, every adult may breed. There are many different and highly successful ways of being a social mammal.

Here is the picture of beyond-baby caring: The expanding circles of caring characteristic of humans and many other mammals—extending beyond an individual and her offspring to include *my* mate, *my* kin, and *my* affiliates—is the expanding of the magic circle of *me and mine*. It entails widening the range of individuals who you care about. These valued individuals may, in consequence, receive aid, defense, comfort, or warmth, depending on what the conditions call for and what can be given. But the pain of another's distress and the motivation to care seems to fall off with social distance. By and large, motivation to care seems to be stronger for offspring than for affiliates, for friends than for strangers, for mates than for friends, and so on. Familiarity, kinship, and kithship (being part of the group) are typically important factors in strength of attachment. There are, of course, exceptions, for in biology, almost nothing is universal. There are some people who act more generously toward strangers than to their own family.

If the maternalization of the brain means that care extends to offspring via mechanisms in the hypothalamus, what mechanisms are used to extend care to mates and others? The darlings of this story are the prairie voles (*Microtus ochrogaster*), who tend to bond to their mates for life. Voles are chubby little rodents with short tails, and there are many species. Bonding means that mates prefer each other to any other vole and show stress when separated. In prairie voles, permanent bonding occurs after the first mating. Male prairie voles also guard the nest and participate in rearing the pups. Bonding does not imply total sexual exclusivity, as it turns out, but regardless of the occasional dalliance, the pair remains as mates. They hang out with each other, and prefer each other's company to that of other adult voles.

Montane voles, by contrast, do not exhibit comparable social behavior, nor does the male have any role in guarding the nest or rearing the pups. They are not social and do not like to huddle or hang out with each other. The mother nurses the pups, but for a shorter period than does the mother prairie vole.

Because these two species are so very similar, other than in their social behavior, Sue Carter, a neuroscientist who studied the sociality of prairie voles, pursued this question: What are the differences between the brains of prairie voles and montane voles? Getting the answer involved the inevitable dead ends and devotion to designing experiments to reveal answers. It turns out that the differences do not involve a whole new chunk of tissue (not macrostructural). Rather, the differences are microstructural and somewhat subtle, pertaining mainly to oxytocin, vasopressin, and extra receptors on neuronal membranes that bind those hormones and thereby affect neuronal activity.

In one very particular region of the reward system (the *nucleus accumbens*), the prairie voles contrast with the montane voles in having a higher density of receptors for oxytocin. In another very particular region of the reward system (the *ventral pallidum*), prairie voles have a higher density of receptors for vasopressin. It should also be noted that oxytocin is more abundant in females than in males, vice versa for vasopressin. Vasopressin and its receptors seem especially relevant to male behavior, such as aggression against nest intruders and care of the pups.

The differences in receptor density are the main circuit-level differences that explain long-term attachment of mates after the first mating, though undoubtedly there are other factors involved as well. For example, after mating, the mates need to be able to recognize each other as individuals. Recognition requires learning, which is mediated by the neurotransmitter dopamine. So if you block the receptors for dopamine, the voles cannot remember who it was they mated with, and so tight bonding with a particular mate does not occur.

Though very common among birds, strong mate preference is somewhat uncommon in mammals. Only about 3 percent of mammals, including prairie voles, pine voles, California deer mice, beavers, titi monkeys, and marmosets, show mate attachment.[15] For these mammals, caring and valuing extend not only to offspring, but to mates: they prefer to be with their mate rather than another adult, and males typically help to rear the offspring and guard the nest. They show anxiety when separated and relief when reunited. In prairie voles and marmosets, but not in rats, attachment extends even to mature offspring. Also noteworthy is the fact that these mature, independent offspring may help their parents feed and tend the next litter of baby prairie voles. This is alloparenting.

When you feel fear or distress, when you feel love and contentment, the major brain players generating these emotions are in the limbic system, which includes the cingulate cortex and the hypothalamus (Figure 4.5). And the limbic system is also the theater for oxytocin and vasopressin action. How exactly do oxytocin and vasopressin see to it that you care for others? A proper answer would involve the details of all the relevant circuitry and how the neurons in the circuits behave. Unfortunately, these details are not yet known. What is known is that oxytocin downregulates the activity of neurons in a structure mediating fear responses and avoidance learning—the amygdala. Oxytocin, as noted, is a kind of safety signal: "Things in my social world are fine."[16] Is oxytocin the *love molecule* or the *cuddle molecule*, as has sometimes been suggested? No. The serious research on oxytocin reveals how very complicated its action is and how complicated is the circuitry underlying social attachment.[17] Some of the early claims relating differences in the gene for the oxytocin receptor (OXTR) and social preferences have not been borne out,[18] and some remarkable claims about correlations between strength of love and blood levels of oxytocin are so astonishing as to raise a red flag regarding experimental procedures. Caution is in order.

When animals are in high alert against danger, when they are preparing to fight or flee, stress hormones are high and oxytocin levels are low. When the threat has passed and you are among friends, hugging and chatting, stress hormones back off and oxytocin levels surge. So not only are the amygdala-dependent fear responses calmed, but the brainstem switches from fight-and-flight preparation to rest-and-digest mode.

In all social animals, shunning and isolation are a form of punishment. That is basically because in shunning, oxytocin levels fall and stress hormone levels rise. Inclusion and touching are, by contrast, sources of pleasure and sources of ease. The insula area of the cortex, known to monitor the physiological state of the body, also helps create the "emotional safety" feelings.

Lest it be thought that if something is good, more of it will be better, here is a cautionary tale. If you inject extra oxytocin into a happily mated female prairie vole, her degree of mate attachment actually wanes, not rises, and she may become promiscuous. What's more, a dose of oxytocin injected into the brain of a female adolescent vole is likely to bring about ovulation and preparedness to mate. We are not sure what it might do to a human female, though it is wise to remember that oxytocin is a very potent hormone. So quick ideas about spraying oxytocin liberally around the playground or the congressional chambers should be reconsidered. The *Goldilocks effect* is seen in many areas of biology: Too little of something is not good, too much is not good. It should be *juuuust* right. Balance, Aristotle's watchword, is the key to biological well-being.[19]

ARE HUMANS MONOGAMOUS?

WHAT ABOUT *human* mate attachment? Are we, by nature, like prairie voles? The answer seems to be that humans are flexible in their mating arrangements. Strong attachments are certainly common, but according to anthropologists, about 83 percent of societies allow polygynous (more than one wife) patterns of marriage.[20] Depending on conditions, even if polygyny is allowed, most men are of modest means and hence likely to have only one wife.[21] Consequently, de facto monogamy may prevail, though the wealthier men may have more than one wife. In historical times, it has been well documented that a wealthy man may have a special, long-term attachment to one particular female, even while enjoying, and perhaps impregnating, various other women in his harem. So nothing much can be inferred about attachments even when the local customs permit polygyny. In the other 17 percent of societies, both modern and ancient (for example, Greece and Rome), monogamy has been the practice.

The explanation for the cultural variation of marriage practices probably rests mainly with variation in ecological and cultural conditions—in particular, with whether there are conventions for the heritability of property and other forms of wealth, along with the existence of wealth to be inherited.

Drawing on historical and ethnographic data, evolutionary biologists[22] argue that when there are multiple wives, each with children and hence multiple heirs, transferring resources to all heirs results in a depletion of the resources. For example, the patches of land to bequeath get smaller and smaller and less able to support the families that depend on the land. A man might select one particular wife whose children inherit all the wealth, but this it apt to make for unpleasant competition among offspring and is generally an unstable solution. In these conditions, a more stable strategy for enhancing the well-being of one's own offspring would be to have one wife, be sure of the paternity of the offspring, and invest heavily in the welfare of her children only.

Monogamy appears to have emerged in Eurasia as agriculture became widespread, with land and herds an important source of wealth that could be passed to heirs.[23] Once certain practices become the norm, once they are seen to bring benefits and to circumvent troubles, once they are reinforced by social approval and disapproval, they are felt to reflect the only right way for things to be. Our intuitions about what is right are strongly shaped by the prevailing conventions.

WHERE DOES MORALITY COME INTO THE STORY?

THE FOREGOING constitutes a very brief overview of how oxytocin and vasopressin operate in the brain to create a platform for sociality and hence for morality. But how do we get from a general disposition to care about others to specific moral actions, such as telling the truth, respecting the goods of others, and keeping promises? How do we get from familial caring to broader community-wide values such as honesty, loyalty, and courage? The answer has two intertwined parts: learning by the young and problem solving by everyone.

In group-living species such as humans, lemurs, and baboons, learning the local conventions and the personality traits of individuals, knowing who is related to whom, and avoiding blackening one's own reputation become increasingly important.[24] Learning, especially by imitation, is the mammalian trick that gets us both flexibility and well-honed skills. Your brain is organized to take pleasure in company and to find social exclusion painful. Your brain is organized to feel bad if your child is maimed or your mate is assaulted. If your brain were organized like that of a turtle, you would spend years alone, entirely contented. Nor would you care if a neighboring turtle got made into soup or baby turtles got eaten by seagulls.

But you have a mammalian brain, and you do care. Human brains, like the brains of baboons and wolves, have the capacity to learn how to get on in a social world as well as in the physical world. You can learn how to avoid an irascible uncle, how to temper your own impulse to strike out, how to compromise, and how to reconcile after conflict. You learn how to bury the hatchet and when to turn the other cheek. You learn to keep promises and tell the truth. You get approval for doing so, disapproval for not. You learn when aggression is called for and when it is not; when help must be given and when it is best to withhold help. Learning these things is possible because your brain is a social brain. Your values are what they are because your brain is what *it* is.[25]

When you were a baby, you began to learn to interact with parents and siblings and to derive pleasure from their company. As you grew up, you automatically adopted their ways of being social, largely without having it explained to you. You imitated, and you developed habits reflecting those manners and practices approved of in your group. You developed a conscience about what is okay and what is not. The reward system of your brain got tuned up so that you feel bad when you steal or even contemplate stealing; you feel good when you resist stealing.

Stories gave you a sense of the right ways to act in a wider context; the ant and the grasshopper, the foolish man who tried to please everyone, the advantages of building a solid brick house rather than a flimsy house of straw. Then there is *The Cat in the Hat*, who skates a bit close to the edge, who does things he should not. There are Harry Potter and Tom Swift and Nancy Drew. Family stories and village stories round out the child's gathering sense of thrift and caution, of justice and honesty.[26]

You observed, sometimes quite automatically and implicitly, sometimes explicitly and with reflection, the advantages of cooperation. Two kids rowing a boat gets you across the lake much faster; two kids turning the long skipping rope allows doubles skipping; turn-taking means everyone gets a chance, so the games do not break down. A group of people working together can raise a barn in one day. Singing in a group with parts makes beautiful music. Pitching a tent is easier with two people, and hiking together provides safety. A child comes to recognize the value of cooperation.

This does not mean that there is a gene for cooperation. If you are sociable and you want to achieve something, then a cooperative tactic can seem a fairly obvious solution to a practical problem. Your big prefrontal brain figures this out fairly quickly. As philosopher David Hume observed, a crucial part of your socialization as a child is that you come to recognize the value of social practices such as cooperation and keeping promises. This means that you are then willing to sacrifice something when it is necessary to keep those practices stable in the long run. You may not actually articulate the value of such social practices. Your knowledge of their value may even be largely unconscious. But the value shapes your behavior nonetheless.

What I call *problem solving* is part of your general capacity to do smart things and to respond flexibly and productively to new circumstances. When you are described as reasonable, in part that means you are a good problem solver and that you can use your common sense. Social problem solving is directed toward finding suitable ways to cope with challenges such as instability, conflict, cheating, catastrophe, and resource scarcity. It is probably an extension to the social domain of a broader capacity for problem solving in the physical world. Depending on what you pay most attention to, you may be more skilled in the social domain or in the nonsocial domain. From this perspective, moral problem solving is, in its turn, a special instance of social problem solving.[27]

Social problem solving is not always about formulating rules, but frequently concerns when or whether an existing rule applies to a case at hand. The co-op hardware store in our village supplied farm equipment, ladders, picking bags, poultry feed, and other essentials to the farmers. It was, in its humble way, a truly farmer-owned and efficiently run business. Members of the co-op elected a board to oversee the management of the store. Dividends, when there were any, were paid to members, and the supplies were modestly discounted for members. The manager was paid a salary, as was the accountant, Aubrey Crabtree.[28] This is a story about the accountant, who after some years of seemingly fine work was discovered to have embezzled what was then, and for these farmers, a princely sum. In today's values, it was roughly equivalent to $250,000.[29] The sum was vastly more than our farm was worth. Because it was co-op money, it was actually money out of our family purse, a fact not lost on me when I compared the Crabtree car with our car, about 15 years old and kept roadworthy by constant tinkering by my father.

Do the existing rules—the provisions of the criminal law—apply to this case? The problem was, he was charming, witty, and likable. He was a great favorite about town. Crabtree was prominent in the theater club that staged small productions once or twice a year (this event happened well before television became available on our area). He was a highly visible and respected member of the Church that the English, as opposed to the Scots, attended, and he sang a strong baritone in the church choir. Inspiring much mirth, he also was in charge of handing around the collection plate.

With the revelation that Crabtree had embezzled, a problem arose because some of his English friends, also prominent in the community, wanted to handle the business locally and in their own fashion, avoiding the whole distasteful involvement of the law and criminal charges. Pressing charges, it was argued, was both unnecessary and unseemly. To aid their case, Mr. Crabtree made a public display in his church of his heartfelt repentance, complete with elaborate apologies and copious tears.

The Scots (or "Scotch," as we said) felt that this was the sort of case for which Canada's criminal law had been thoughtfully crafted.[30] They argued that the law needed to be respected, friendship notwithstanding. It was arrogance to suppose that friends could make a decision that would trump the wisdom embodied in a well-tempered, long-standing criminal code. Having politely listened to Crabtree's repentance, they suspected that thespian skills were more in evidence than genuine regret. After all, the theft had taken place stealthily and methodically over a period of five years by the very person who was entrusted with doing the books, not raiding the books. Perhaps his main regret was having been caught.

For those of us in our mid-teens, this was an absorbing problem. This was real and present. It was not a textbook case. It concerned a person we all knew and liked. It raised painful questions about special treatment for friends who violated the law, about when a local magistrate should give an otherwise decent person a second chance, about whether a planned crime can be excused as an error in judgment, about the effect on deterrence in general if the law were not systematically followed, about whether restoration of funds and public humiliation were "punishment enough." Every one of these issues emerged in essentially the same form when, as a professor, I saw faculty called upon to confront issues of academic misconduct. "Let's just handle this ourselves; no authorities need to be involved; we are best positioned to judge the authenticity of expressed regret; it is unseemly to punish a good friend; this is our tribe," and so forth. That I had seen the pattern before helped me to think through these academic cases.[31]

As for the Crabtree case, at supper tables on many farms, heated discussions took place concerning the fact that lesser crimes by the aboriginals from the NK'MP (pronounced "in-ka-meep") reservation were indeed brought to the attention of the law. Because Crabtree had children in high school, we were also made aware of the moral requirement to treat the children with normal respect, never to tar them with the brush that tarred their father. By and large, so far as I could tell, this was observed, though doubtless they did suffer acute embarrassment privately.

In the end, Crabtree was given a reprieve. The co-op board voted against bringing criminal charges, taking on good faith his promise to repay. My father, one of two dissenters in the vote, resigned from the co-op board, dolefully warning that most of the money would never be repaid. Sadly, he was right, and there was no great comfort in that. Only a small fraction of the money was returned, and the Crabtrees left the valley to start fresh elsewhere. Embarrassed by Crabtree's dereliction, his defenders made the very topic unacceptable in polite company. This, too, was a pattern I saw later in academic life, even among philosophers who made studying morality their domain of expertise. They did not want to revisit their "let's do it in-house" decisions.

Although evaluating how to proceed with a particular case is frequently the most pressing concern, the more fundamental problem concerns general principles and institutional structures that undergird well-being and stability. The development of certain practices as normative—as the right way to handle *this* problem—is critical in a group's cultural evolution.[32] There are established principles enjoining group members against such behavior as embezzlement and other forms of cheating. Motivated to belong, and recognizing the benefits of belonging, humans and other highly social animals find ways to get along, despite tension, irritation, and annoyance. Social practices may differ from one group to another. The Inuit of the Arctic will have solved some social problems differently from the Pirahã of the Amazon basin in Brazil, if only because social problems are not isolated from physical constraints such as climate and food resources.

Similarities in social practices are not uncommon, as different cultures hit upon similar solutions to particular problems. This is akin to common themes in other practices, such as boatbuilding or animal husbandry. Particular cultures developed skills for building particular styles of boats—dugout canoes, birchbark canoes, skin-backed kayaks, rafts with sails, junks for fishing on the rivers, or what have you. After many generations, the boats made by separate groups are exquisitely suited to the particular nature of the waters to be traveled on and the materials available. Ocean or lake or river? Stormy or calm? Gentle breezes or powerful prevailing winds? Notice, too, that many different cultures learned to use the stars for navigation. Some picked up the trick from travelers, others figured it out independently, just as conventions for private property occurred in different groups as their size expanded and as agricultural practices became widespread. I am reasonably confident that there is no gene for navigating by the stars or for building boats.

The domestication of milk-producing ungulates such as camels, cows, and goats is another example of widely separated cultures hitting upon similar solutions to a resource problem. Tethering animals or building corrals is also common, but reflects a similar result of human problem-solving capacities. Camels are not available in Ireland, so not surprisingly, goats were domesticated there. Neither goats nor camels are available in the Arctic, so domestication of ungulates did not happen there. Until very recently, Inuit groups remained very small owing to resource scarcity, and hence there was no pressure for community-wide institutions of criminal justice. For the Inuit, judicial needs were infrequent and were met informally by the elders.[33]

Though expressions of moral values vary across cultures, they are not arbitrary in the way that conventions for funerals or weddings tend to be. Matters of etiquette, though important for smoothing social interactions, are not as serious and momentous as moral values. Truth telling and promise keeping are socially desirable in all cultures. Is there a gene for these behaviors? Though that cannot be ruled out, there is no evidence for a truth-telling or a promise-keeping gene. More likely, practices for truth telling and promise keeping developed in much the same way as practices for boatbuilding. They reflected the local ecology and are a fairly obvious solution to a common social problem.[34]

A philosophical objection to the approach outlined here is that social behavior anchored by the brainstem-limbic system and shaped by reward-based learning and problem solving cannot be genuinely moral behavior. That is because to be genuinely moral, the behavior must be grounded solely in consciously acknowledged reasons concerning, say, the well-being of others. According to this view, to do the moral thing, you must do it only *because* it is the moral thing, and the action must be an application of an absolute law. This rules out the moral behavior of most people living in hunting or gathering groups as not genuinely moral. I suspect that these very stringent criteria entail that you and I are not moral agents either. Which implies that maybe the criteria are preposterously stringent, dreamed up in the armchair, far from the realities of the lived social world.[35]

One response to this philosophical objection is that any account of the basis for morality needs to be neurobiologically, anthropologically, and psychologically realistic. Certainly, reasoning—or more generally, thinking and problem solving—is a highly important part of social life and moral life in particular. Nevertheless, thinking and problem solving in the social domain is constrained—by our past learning and intuitions, by our emotions, and by the way our brain is tuned to the needs of kith and kin. Problem solving involves many factors and cannot be reduced to something like a syllogism. Reasoning, as we shall see in Chapter 8, is probably a constraint satisfaction process, whereby the brain weighs and evaluates and considers a host of factors and settles on a satisfactory decision.

A different sort of problem arises with regard to one particular aspect of morality, namely, that concerning fairness. Is recognition of what is fair or unfair also anchored to the same platform as other moral dispositions, such as cooperation and truth telling? One line of evidence suggests that fairness may be different. Ethologists Frans de Waal and Sarah Brosnan showed that capuchin monkeys are acutely sensitive to what another monkey gets as a reward. The monkeys are caged so that each clearly can see the other. If monkeys A and B both get a cucumber slice, there is no problem. They enjoy their cucumber slices, but prefer grapes, as do I. So if monkey A gets a cucumber slice and monkey B gets a grape, monkey A swiftly and accurately throws his cucumber slice at the experimenter and then angrily rattles the cage bars. If this happens the next time, monkey A is totally furious and not only throws the cucumber at the experimenter, but beats on the cage, slaps his hand on the floor outside the cage, and vocalizes his displeasure. The monkey recognizes unfairness. Note that the capuchins like cucumbers just fine, but they much prefer grapes, so it is not that they were given something they hate or is inedible.

In a slightly different condition, sensitivity to getting the short end is again manifested. If monkeys A and B have to "work" for a reward— pass a small stone from a bowl inside the cage to the experimenter, for example—and they get unequal rewards, again the poorly rewarded animal is outraged.[36] This behavior of the capuchins clearly shows recognition of fairness with respect to oneself. As Sarah Brosnan reports, however, the capuchins do not object when they are the beneficiary and do not alter their behavior to get food for their partner as well as for themselves.[37] A reasonable guess is that the capacity for evaluating fairness arose from within-group competition for resources. As Brosnan points out, what the capuchins seem to show is not so much concern that others get fair treatment but concern for oneself—*I should get what he gets*. And of course, this is a strong trait in humans.

Owen Flanagan has suggested to me that such self-oriented evaluations of fairness appear to be independent of caring and attachment.[38] This raises the possibility that fairness, as an aspect of social behavior seen in humans, for example, may not be linked to the same mechanisms as helping, sharing, or cooperating. So the question regarding the platform for the moral value of fairness is this: How does the brain get from be-fair-to-*me* to concern for the fair treatment of others—that is, be-fair-to-*him*? Is there a role for attachment circuitry?

Caring about the fair treatment of those to whom we are attached is plausibly linked to the extension of self-care to care for offspring, mates, and friends, especially in those mammals that have the capacity to see things from the vantage point of others. Children often show concern that their siblings are treated equitably when gifts are distributed and feel bad if a sibling is left out. This seems especially true of an older sibling with regard to the younger ones. I think this response is anchored by the caring platform.

Not always or everywhere, however, is equitable treatment for all a norm. Inequitable treatment of males and females, parents and children, slaves and free persons, the rich and the poor, officers and enlisted men, teachers and pupils, and so forth, is very common. Some inequitable practices, such as those relevant to parents and children, are pragmatically defensible, however much the children may sometimes holler. Evidently, application of fairness as a moral category is, in practice, highly variable with lots and lots of places where people, even within a culture or a family, simply disagree. For example, well-meaning, morally upstanding, conscientious, decent people can disagree about whether affirmative action in college admissions is fair or unfair; they can disagree about whether a flat tax is more fair than a graduated income tax or whether inheritance taxes are fair. Sometimes the charge "that's unfair!" is not much more than an expression of opposition and a demand for change that exploits our fast response that unfairness is wrong.

Being reminded of the variability in what counts as fair helps us acknowledge that standards of fairness are not universal, that they can change in unpredictable ways. More generally, it reminds us that moral truths and laws do not reside in Plato's heaven to be accessed by pure reason. It reminds us that fairness calls are often mixed with a whole range of emotions, including fear, resentment, empathy, and compassion.[39] The idea of universal human rights, laudable as it is, emerged very recently and may be linked to the expectation of the generally beneficial effects of extending fair treatment, especially in the legal domain, to all.[40] Thus, when philosophers or psychologists claim that we humans are all born with an innate module to behave according to fairness norms, we should wonder how they square such a hypothesis with the aforementioned variability.

WHEREFORE RELIGION AND MORALITY?

MY AIM in this chapter has been to look to neurobiology and the evolution of the brain to try to understand the platform for moral norms and conventions. This approach addresses only the basic motivations and dispositions that enable social behavior, not the specific character of the norms adopted by a group. Two important questions remain: (1) What is the role of religion in morality, and (2) what about the dark side of human nature—our tendency to hate those in the outgroup, to kill, assault, maim, and cause mayhem? First, religion.

Organized religions began to emerge at about the same time that agriculture took hold as an important way of getting food—about 10,000 years ago. This is very recent in the history of *Homo sapiens*, since we have been on the planet for about 250,000 years. Other hominins, such as *Homo erectus*, appeared about 1.6 million years ago. Although we know nothing of the social lives of *Homo erectus*, it would not be unreasonable to suspect that they lived in small groups and had many of the same basic norms typical of existing hunter-gatherer societies, such as those in the Amazon basin.[41]

For the greatest segment of our history (about 240,000 years), our social and moral lives were conducted without beliefs in law-giving deities and an organized priest class. Anthropological data gathered over the last several hundred years have helped us understand the social lives of people with cultures very different from the cultures of humans living in cities in Europe and North America. The Inuit, who were primarily hunters and fishers, had very thoughtful, practical norms governing their social lives. Interestingly, deceit was considered especially wrong, even more so than murder. Whereas murder was rare, usually involved a woman, and might affect only one person, deceit could imperil the whole group.

A basic postulate underlying Inuit culture was that spirit beings and the souls of all animals have an emotional intelligence similar to that of humans.[42] This reflects the nature of their dependence on and interaction with the animal world—with polar bears who are tricky to outwit, with devious seals who can escape capture, and with whales that are so awesome that I find it hard not to believe they are deities of some kind. As the legal scholar and anthropologist E. Adamson Hoebel reported, the Inuit usually did not *explicitly* formulate their norms and practices. Norms and conventions were simply picked up by children from those around them, much as they picked up the skills needed for kayaking, sealing, and whaling. Thus, most Inuit knew, without being told, that aggressive behavior must be kept within strict limits; it can be displayed in a controlled manner in specific games, but should be suppressed or redirected in ordinary life.

For a different example, studies indicate that farming groups, such as the Ifugao (in the northern mountains of the Philippines),[43] also had thoughtful moral practices, including those for settling disputes with the help of a mediator, specifying grounds for divorce, for borrowing and lending items, and for transferring property. By necessity, they had enforced rules governing irrigation—important because they cultivated rice in terraces on steep slopes and had to share a water system. Some of Ifugao practices are surprising. For example, there was no chieftain, only a governing council. Also, they assumed that the bilateral kinship group is the primary social and legal unit. It consists of the dead, the living, and the yet unborn.[44] Spirits of one sort or another seemed to be invoked, and sacrifices were thought to be useful in getting along with the spirits, but no concept even approximating the Judeo-Christian God was part of their culture.

Organized religion, so much a feature of contemporary life, became a feature of human culture only recently. From my perspective, therefore, it is not that organized religions invented and set up moral rules as much as they adopted and modified the traditional moral norms that were the implicit practices typical of small human groups. Organized religions, like the implicit practices of people such as the Inuit, depend on the neurobiological platform for sociality that has been part of human nature all along.

Roughly within the last 10,000 years of *Homo sapiens'* approximately 250,000-year existence on the planet, human cultures have developed in uniquely rich ways. Consequently, cultural and social institutions have changed the ecological conditions for modern humans, whose social lives are very different from those of humans who lived 200,000 years ago. Groups have grown large, and interaction has expanded far beyond the occasional gatherings where small groups could meet and exchange tools. "Niche construction" by humans has changed the species' ecology in many ways, altering social organizations and leading to the formulation of specific laws and rules. Social behavior—including what we would now identify as moral behavior—has changed accordingly. One way in which some communities have changed concerns organized religion.

Not surprisingly, when groups become so large that not everyone knows everyone else, the power of one-on-one disapproval is much diminished. Formulating rules explicitly to govern everyone is part of a solution to the social problem of compliance. In some large groups, a local wise elder may have found himself in charge of specifying the rules and seeing that they were properly enforced.[45] In other groups, this task may have come to be assumed by the prevailing shaman. The idea of an invisible deity that can see you no matter where you are is perhaps helpful in deterring antisocial behavior that might otherwise be thought to be undetectable. The efficacy of this postulate in achieving compliance is difficult to measure, however.[46]

Not all religions associate the source of morality with a god or the teachings of the clergy. Buddhism, Taoism, and Confucianism are three such examples, and these are neither small nor inconsequential religions, needless to say. Many so-called *folk religions* endowed animals such as ravens and eagles with special status and saw physical events such as seasons, the tides, storms, and volcanic eruptions as involving spirits of one sort or another. A supreme lawgiver, however, was something more typical of religions that became dominant in the Middle East. It is not a universal feature of religions.

The history and cultural development of religions, and monotheistic religions in particular, have been well documented elsewhere, and it is safe to say that monotheism is a development from earlier religions that had assorted gods and spirits, many with very human-like virtues and vices.[47] My main point is that moral behavior and moral norms do not require religions. Nonetheless, a religion may add to existing norms or create completely new ones, such as requiring animal sacrifices on particular days of the year or mandating particular dietary or clothing rules. These norms are often prized for highlighting the differences between groups, with special focus on those demarcations that are highly visible. Us versus *them*. Religions may also provide a forum for discussing moral dilemmas and difficulties, for reinforcing group norms, and for stirring up the appropriate emotions in the event of battle.[48]

TENSIONS AND BALANCE

THE MAMMALIAN BRAIN is wired both for self-care and for care of others, and on many occasions, the two are not harmonious. Social life brings benefits, but it also brings tensions. You compete with siblings and friends for resources and status; you also need to cooperate with them. Some individuals are temperamentally more problematic than others. In a very large group, there is apt to be less homogeneity in how children are brought up, and hence socialization patterns can vary a great deal. Some children are encouraged to be self-centered, others are encouraged to take in the other's point of view. Sometimes you have to tolerate people who are irritating or noisy or smelly.

Even though you love them, your children, your spouse, or your parents can provoke you to anger. You can find them disappointing and frustrating in all kinds of ways, major and minor. You can also find them uplifting, loving, and wonderful in all manner of ways. Sometimes tough love serves the child better than indulgence; sometimes turning a blind eye and a deaf ear is the sensible course of action. None of this is news. It is just a reminder of what you already know and live with all the time. Your social life has both pains and joys and has them at the same time. You have self-care circuitry and other-care circuitry, and the two do not always line up.

Social life can be very subtle, often calling for wise judgment rather than strict adherence to rules. As Aristotle as well as the Chinese philosopher Mencius well realized, the drawback with rules is that there cannot be a rule to cover every contingency or for every situation that may crop up in life. Rules are just general indicators, not hidebound requirements. Judgment is essential. Moreover, even for general norms that usually apply, exceptions may have to be made if conditions call for them. Sometimes telling a lie *is* the right thing to do if it saves the group from a madman threatening to blow up a bomb, for example. Sometimes breaking a promise *is* the right thing to do if it prevents a truly terrible catastrophe, such as the meltdown of a nuclear reactor. The problem with exceptions is that there are no rules for determining when something is a legitimate exception to prohibitions such as *don't lie*, *don't break a promise*, and *don't steal*.

Judgment is essential. Children quickly learn about prototypical exceptions and apply fuzzy-bounded categories rather than hidebound rules.[49] You understand that killing another human is wrong but that under some circumstances, such as in self-defense or in war, it is not wrong.

Differences in group history or in the ecological conditions in which different groups live can mean that some cultures put high value on traits such as thrift or modesty, while other cultures value displays of generosity and consider boasting a mark of good character. Had you been born in an Inuit camp in the eighteenth century, you might well have abandoned a neonate female in the snow if you had already given birth to three female children. You would have regarded this as regrettable, but certainly not disgraceful. The ever-present threat of starvation and the need for productive hunters meant that harsh things had to be done.

Though it may be urged upon us to treat everyone as equally entitled to our care, typically this is not psychologically possible, nor, I think, morally desirable. Typically, your own children's welfare will come before that of unknown children living on the other side of the planet; sometimes charity does begin at home, and it would be wrong to abandon your ill parent to tend to a clinic full of lepers in Nepal. In fact, you are generally not expected to care impartially for all and sundry; you apportion your caring, you balance your generosity. Balance, as all wise moral philosophers have emphasized, may not be precisely definable, but it is needed to lead a good social and moral life. Not every beggar can be brought home and fed, not all your kidneys can be donated, not every disappointment can be remedied.[50] But we cannot codify balance in an explicit rule. Balance requires judgment.

The tensions of social life wax and wane, and individuals often figure out small ways to organize their lives to reduce unavoidable tension. But you cannot expect tensions to disappear altogether. Often there is no perfect solution; not everyone can agree, nor even agree to disagree. Sometimes there is no solution at all beyond accepting what is and carrying on. Sometimes social practices can exert such a hold on emotions that they can overpower the attachment to offspring.

The place is Mississauga, Ontario, the year 2007. Aqsa Parvez, at age 16, wished to no longer wear the hijab. She quarreled with her family over the issue. She was strangled in her home in December of that year. None of her other siblings or her mother helped her. Her father and brother pleaded guilty to second-degree murder and received life sentences. She had gone to her first movie shortly before her death. There have been 15 such "honor" killings in Canada since 2002.

———————

MORAL NORMS and conventions are shaped by four interlocking brain processes: (1) *caring* (rooted in attachment to kith and kin and care for their well-being; (2) *recognition of others' psychological states*, such as being in pain or being angry; (3) *learning social practices* by positive and negative reinforcement, by imitation, by trial and error, by analogy, and by being told; and (4) *problem solving in a social context*, for example, as new conditions produce new problems about resource distribution and migration, about interactions with outgroups, about how to resolve disputes about ownership of land, tools, and water.[51]

What I have left to the side in this chapter is aggression and hatred, very strong forces in social life. In the next chapter, we shall explore further the potency and lure of hatred.

Chapter 5

Aggression and Sex

THE JOY OF HATING

SOMETIMES PLAY FIGHTING crosses the line into real fighting. Sometimes defensive combat emerges when trust should prevail. Sometimes the wiring for impulse control is overwhelmed. By ideology. By rhetoric. By fear and hate. Sometimes . . . all hell breaks loose.

Fans of the San Diego Chargers football team are full of hate for fans of another California football team, the Oakland Raiders. They taunt each other, donning costumes to intimidate or humiliate the opposing team's fans. Some of the fans engage in ritual fight displays, not unlike those of aggressive birds such as the Noisy Miner.

Chargers fans say the Raiders are evil, disgusting, and subhuman. And vice versa. Maybe it is just play hate. Evidently it *is* fun. All sides hugely enjoy the hate fest. Any casual observer can see that the fans derive enormous pleasure from belonging to a group that is united in its hate for the other group. The very hate itself seems to be exciting, invigorating, and pleasurable.

Not incidentally, an astonishing amount of time and expense goes into this ritualistic hostility.

Nevertheless, in the United States, fighting between the fans at football games (U.S. football) is quite rare. On the exceptional occasion when it does occur, fans generally express horror and outrage. In England, however, one group of fans having it out with another group is not rare. Fighting among male fans after matches, and sometimes during and before matches, has been disturbingly popular among a subset of men. Football brawls happen routinely. Hooliganism has been exceptionally difficult to wipe out.

The BBC documentary on football fight clubs shows that for many young men, brawls between rival groups are terrifically exciting. Brawls are a major reason for attending matches, whether in the hometown or in France, Italy, or elsewhere in Europe. In the United Kingdom, the gangs are referred to as "firms." Football firms are well organized, with a "top lad" who plays a leadership role and organizes fighting events around football matches.

What are the men of football firms like? To judge from the BBC documentary, they are charming, articulate, and bright. They were not beating their chests or frothing at the mouth. They did not look crazed. They could be your brother or cousin. The normality of their manner seems incompatible with their love of brawling, and yet it is not. This is essentially brawling without cause. It is brawling for the sheer fun of it.

The Los Angeles riots of 1992 erupted as racial and ethnic tension boiled over following the acquittal of three white policemen who had been videotaped viciously beating a black man, Rodney King.[1] The outrage at the unfairness was profound, and suddenly all hell was unleashed: arson, looting, and shooting were occurring all over South Central Los Angeles.

I saw the video images of the hapless white trucker, Reginald Denny, forcibly yanked from his truck at an intersection by four black youths. They savagely kicked him and smashed his head with a brick, almost killing him. Watching the event on video again now, I cannot but be stunned, as reporters were at the time, by the joyous body language of the youths as Denny lay semiconscious on the ground. They danced with joy. Mostly unconcerned, people were milling all around the intersection. Fortunately, four black citizens, having seen the Denny beating on television, went to his rescue and took him to the hospital. But for their kind actions, he likely would have died.

Utter chaos reigned in the city for several days. The police had to back off because they were so deeply mistrusted and hated that they had become a popular target of gunfire. The National Guard had held back because their ammunition had not been delivered. Some Korean shopkeepers tried to defend their property by shooting looters, while others elected instead to simply watch as their shops were looted and burned.[2]

Here, in the midst of the frustration and anger of the rioters and looters, there was joy and some sense of justified pleasure in striking back. One woman with a video camera reported, "When I was on the streets, people were having a ball. They were stealing and laughing and having a bunch of fun." At least 54 people were killed, and thousands more were injured.[3]

With absolutely no pretext save losing the Stanley Cup to the Boston Bruins in a fair hockey series, hockey fans in June 2011 went on a rampage in downtown Vancouver, burning cars, looting shops, and causing mayhem. Yes, this was Canada, where these things are not supposed to happen. In this instance, too, the joy of the fans, mostly young men, was unmistakable. They danced on overturned cars, smashed store windows, set fire to vehicles, and taunted the police, who struggled to maintain some semblance of order.[4]

Cage fighting, I am told by my friend Jonathan Gottschall, is his passion. This seems peculiar to me, as he is a professor in a literature department.[5] He tells me that the appeal of cage fighting is completely different from that enjoyed in riots and brawls. As Jonathan says, this is basically a friendly form of mutual assault. One-on-one, where the fight is fair, in the sense that officials match the pugilists in age, weight, and standing, is very different from mob brawls. Fear is the overriding emotion before the fight; intense attention is the main mental state during the fight. According to the cage fighters, the only pleasure comes at the end of the fight, and only if you win. The pleasure of defeating your opponent is so incredibly intense, it makes the risk worthwhile. Some cage fighters say it is a kind of ecstasy, comparable only to sex. This connection is less surprising than it first seems. Sexual behavior and violence are linked in the brain, in a region of the hypothalamus (ventral medial). In male mice, activation of some neurons in this tiny area provokes aggressive behavior to other males put in the cage, but provokes mating behavior when females enter the cage.[6]

Hate gets classified as a negative emotion, and we might assume that negative emotions are the opposite of pleasurable. But in reflecting on the hate of sports fans or rival gangs, you cannot but notice that it tends to be energizing. Arousal is pleasurable.[7] Sometimes it is called being "adrenalized." So it is.

The comedian Louis C.K. describes standing in a long queue at the post office. He looks at other people in the line. He immediately sees things about them to hate. What idiotic shoes that guy wears; what a dumb question the customer is asking; what a loser. Contempt keeps him amused until his turn comes up. Despising others, however trivial the pretext, feels wonderful.

What else is going on in the hate state? You are familiar with guilty pleasures—doing on the sly something that is only modestly bad but still forbidden, like showing each other your bare childhood bums behind the barn, for example. What fun when you are 5. What a delicious secret to keep from your parents. In watching the videos of the Vancouver riots, the joy of breaking the rules, and doing so with others, was palpable.

Women are usually bystanders in hostility rituals and murderous raids. By and large, the perpetrators are men, and mostly, apart from a leader, they are young or middle-aged adults. A long-standing hypothesis is that the males of a gang are, in part, performing for each other. Their hostility displays reassure one another of their mutual attachment, their reliability in case of attack, and their common purpose. Their bonding gives them feelings of power, the power of numbers. That is linked to pleasure. Clothed alike in white sheets, rhythmically dancing around a huge bonfire, the men of the Ku Klux Klan appear to be having a ball. If duty alone were the incentive, no one would show up.

Females are not without aggression, however. Generally, but not always, it just takes a different form. Mean gossip, unkind cuts, shunning—all are potent forms of aggression used by females. Hair pulling seems to be making a comeback these days. Here as well, some form of pleasure seems to derive from collective hating. As with the football fans, there seems to be both intense ingroup bonding and intense hostility to those in a rival group or perhaps someone excluded from all groups.

Us versus them delineates the border of one's safe group. Within the group, individuals can count on affection and adherence to group norms. Outside the group, interactions are riskier and individuals have to be more vigilant. The form that the hostile behavior takes toward those in the outgroup depends on what is in your toolbox, which depends in part on your genes and in part on what you pick up from your culture as the right way to do things. You model yourself after those you admire. Much simplified, in many cultures boys bash, girls shun.

When I was in ninth grade, a rather homely and forlorn girl with whom we had all been acquainted from the first grade began to be visibly pregnant. Being slow in class, she had generally been shunned as "retarded." She failed to pass through the grades and struggled to learn to read. Dorothy was, so far as the girls in their tight cliques were concerned, essentially a nonentity. How had Dorothy become pregnant? She had no boyfriend, and in our village, everyone knew who was dating whom. As we all came quickly to know, one of the local lads, a logger, had taken her out and "had his way with her." Beer was likely a factor. Did we feel sorry for her? Did we offer condolences for what was surely rape or the next thing to it? Were we dismayed by the lad's taking advantage of a simpleminded girl?

Not a bit of it. We wallowed in our superiority, we basked in our wholesomeness and how grand it was that we were not Dorothy. We thought that her predicament was the sort of thing that happened to a girl like her; certainly not to girls like us. Such contempt was not an emotion generated when each of us was alone; then, individually, we were pretty scared by what had happened to Dorothy. Somewhat ignored before, Dorothy now was completely ostracized. Disgraceful though our behavior certainly was, the contempt quickly, and yes, joyfully, bubbled up when three or four of us assembled. Such scorn was part of what kept us so tightly bonded together. Hate binds, and social bonds are a joy.

At about the same time that we scorned Dorothy's predicament, a chore befell me at the farm. Our white leghorn hens, generally a sociable crew, had ganged up on one miserable hen that had somehow acquired a scratch on her neck. The flock would not leave her alone. Swarming around her, the other hens pecked at any sight of blood, opening her wound further. They would have killed her, but my father told me to remove her and make a special pen to put her in until she healed. She still had a year or so of good laying, not yet old enough for the stew pot. Why, I asked, do hens behave in such a horrible way? "Well," came the reply, "I don't know. They just do." The analogy with Dorothy was not lost on me. I regret to say, however, that it made not one whit of difference to my behavior.

Many years later, my friends and I looked back on the Dorothy episode with unequivocal self-loathing. Having matured, we assessed our adolescent behavior with something akin to disbelief. How had we allowed ourselves—even encouraged each other—to be so mean? Could we really have been like that then? Will our daughters be like that?

WIRING FOR AGGRESSION

HOW DOES aggressive behavior serve animals, including us? Ultimately, aggression, in all its themes and variations, is about resources, sustaining life, and passing on one's genes. No surprise there. So, used judiciously, aggressive behavior can often benefit an animal. For predators, aggression against prey brings food. For prey, aggression is defensive, along with fleeing and hiding. Thus, a wolverine is both intensely aggressive in the hunt, but also in protecting food and driving off intruders. Unlike voluntary cooperation in social animals, aggression is very ancient.

If you are engaged in predation or defense, you have to be "up" energetically. You cannot be in "rest-and-digest" mode. As neuroscientist Jaak Panksepp observed many years ago, energy *is* delight.[8] Feeling energetic feels good. It feels exciting.

Consider predators, such as a wolf pack hunting elk. To be effective, the wolves' motivation to kill must be strong, strong enough to overcome fear of a prey that can lethally kick and horn-gouge, but not so strong as to induce recklessness. A delicate balance. The wolves that keep to the rear of the elk try to rip the leg tendons without getting kicked. Those harassing from the front try to keep clear of the dangerous horns. They aim to rip out the throat. The wolves attack the beast and defend themselves at the same time.[9]

This mix of energy, strong desire, fear, and vigilance is crucial for hunting success. Overpowering the prey means success, and success means food. And that, of course, means pleasure. Maybe the pleasure-hostility link is also, in part, owed to the response of the reward systems to a victory, with the result that on the next occasion, mere anticipation of victory brings pleasure. In other words, the animal gets a dopamine hit after the first prey catching. The brain then associates the predatory action with the pleasure of eating. That value is then attached on the next occasion to the goal itself. Anticipatory pleasure is real pleasure.[10] What are the neurochemicals at work in generating this pleasure? Endogenous opioids? Endocannabinoids? Dopamine? All of the above?

Defense against attackers, too, has its "up" feeling, though defense will also involve fear, possibly overwhelming the thrill of being energized for action. Endogenous opioids are probably released, enabling the animal to keep fighting back despite injury. This would partially explain the feelings of pleasure. It also explains the frequent phenomenon whereby only modest pain is felt during a fight, even when a person is shot or hit hard. Success in defense leaves us awash in joyful energy, any fear having utterly vanished. The pleasure of success in beating off a predator is empowering. I can do it![11] Failure? Defeat, especially chronic defeat, makes animals depressed and withdrawn.

Aggressive behavior can also emerge in defense of offspring. From the perspective of evolution, mammalian parents who respond with ferocity to threats to their offspring are likely to have more offspring survive than those who timorously abandon their offspring to an early death—other things being equal, of course. Hence, the genes of the ferocious defenders will spread; those of the timorous will not.

The determination of parents is breathtaking. Crows will mob and dive-bomb anyone who comes close to a newly fledged crow; a mother squirrel will hurl herself into the maw of a dog to give her baby a chance to escape.[12] A mother bear, otherwise fairly shy, can be a powerhouse of fury if she perceives her cubs to be threatened. Protecting the young is a massively powerful impulse among mammals and birds. Human parents may be at their most aggressive in trying to get their child into Harvard or Stanford.[13]

Predation and defense are forms of competition. The intended prey is competing against the predator for its life, the predator is competing against the intended prey for its protein. There is also territorial aggression, which is really a proxy for food. A patch of land will support only a limited number of animals, such as bears or barn owls. The bear who has command of that territory will roust others eager to partake of the berries on his turf. It is easy to see how territorial aggression would have been selected for, just as it is easy to see how predators of one kind or another inevitably emerged. Territoriality probably evolved many times, just as color vision and sociality evolved many times.[14]

Could there have been a world without aggressive behavior? Probably not our world anyhow. Natural selection inevitably involves competition for resources. At some point in biological evolution, some creature will have the capacity to kill and eat others. Just as inevitably, some of those others will eventually be born with the capacity to resist, perhaps by camouflage, perhaps by beating a hasty retreat, making a terrible smell, or fighting back. The arms race is on.[15]

COMPETING FOR THE BEST MATES

THEN THERE is sex. In most mammalian and bird species, there is competition between males for access to females. Such competition can take many forms, but often it entails driving off or outperforming would-be suitors in hopes of impregnating a suitable female. In mating season, in full rut, a couple of bull moose, for example, will fight head-to-head until one finally tires or recognizes his weaker position and scuttles off. In a baboon troop, the alpha male has greater privileges with the troop females than do other males, and he will fight challengers to maintain his status. The lower-ranking males find ways of mating out of sight of the alpha male, sometimes involving significant deception.[16]

Using a different strategy, male bowerbirds build fancy structures to attract females, and competing males will, on the sly, try to wreck them. And among blue Manakins, the males dance, each trying to outperform the other. Females select those perceived by them to have the most sophisticated dance—or so it seems to a human observer. The females select the male, and for these bird species, aggression is less important than performance, presumably because performance has come to signify what the female wants in a mate—virility and competence (a good brain).[17]

Human mating behavior is more complicated. Like much human of behavior, it is subject to such a high degree of shaping by cultural norms and conventions, by fads and trends, that you can only marvel at the flexibility of the human brain. Sometimes human mating behavior seems to have borrowed something from the bowerbird, something from the head-butting moose, and something from the lekking of the sage grouse.[18] In humans, as in most other mammalian species, males compete for females. There are displays, often conventionalized within the norms of a culture: displays of power, strength, wealth, beauty, generosity, cleverness, and social status. Depending on the culture, females may also compete with each other, and female selection may play a major role in mating.

Aggression is multidimensional. It has multiple triggers, variable mixes of emotions, and variable manifestations. It can serve any one of a set of diverse purposes. Within a species, there is apt to be considerable variability among individuals in the taste for aggressive behavior. Being slow to anger may serve animals better in times of peace and plenty, whereas a hot-tempered disposition may be more beneficial during times of scarcity or war. Moreover, an animal may be quick to respond in defense of the brood, but slower in preparation for an attack on prey.

MALE AND FEMALE BRAINS[19]

IN OUR school, fighting was not allowed in the school yard, so the boys who wanted to fight marched over to the footbridge that crossed the irrigation flume, just over the edge of the school's boundary. This was a good venue, as onlookers could seat themselves along the bridge rails and cheer on their favorite. These fights did not amount to much other than the odd bloody nose and a generous helping of crow. The teachers let it go unless the frequency was considered too high, whereupon the principal called in the likely suspects and gave them the strap. That kept the business in bounds. The girls did not fight. They gossiped. They could, of course, be spiteful and cruel, especially to other girls, but they did not actually engage in a slugfest. This tends to be the pattern worldwide, though of course local customs and biological variability allow for exceptions.

It is well known that human males more commonly engage in physical fights than females. The rates of convictions for assault, battery, and murder are vastly higher in men than in women. Men are typically the ones who hold up the stagecoach, rob the banks, go to war, and knife their way out of a bar fight. This is true pretty much everywhere. What explains this difference between males and females?

Testosterone is an essential element in the story, but the story is massively complex in very surprising and interesting ways, making offhand comments about testosterone poisoning in males a poor indication of the facts. Before outlining the role of testosterone in aggression, it may be useful first to briefly address how human male and female brains differ and the mechanisms whereby these differences are established and maintained.

Normally, when a sperm fertilizes an egg, the resulting human conceptus has 23 pairs of chromosomes (Figure 5.1). A pair of sex chromosomes is either XX (genetic female) or XY (genetic male). In the early stages of development, the sex organs (gonads) of the fetus are neutral, but during the second month of fetal development, genes on the Y chromosome produce proteins that transform the neutral gonads into male testes. Absent this action, the gonads grow into ovaries. In the second half of development, testosterone produced by the fetal testes is released into the bloodstream and enters the growing brain. Testosterone now comes to affect the anatomy of the male brain.

How do sex hormones interact with the growing fetal brain? The fast answer is that a surge of testosterone *masculinizes* the fetal brain by altering the number of neurons in very specific areas that mainly concern reproductive behavior, such as mounting and penetrating. In the absence of the testosterone surge, the brain follows a typical female path. So the female brain plan can be thought of as the default—it is what you get unless testosterone masculinizes the brain. As we shall see, however, even when testosterone is available, other factors, including timing of testosterone release and the amount released, may mean that the brain is not masculinized after all or it may be masculinized to some lesser degree.

What does it mean to say that the brain is masculinized? Testosterone affects the number of neurons forming a network. More exactly, it prevents the death of neurons in the affected network. Consequently, several neuronal groups (nuclei) in the hypothalamus are more than twice as large in males as in females.

Why is there any cell death in a *developing* brain? In general, the developmental blueprint for making nervous systems allows

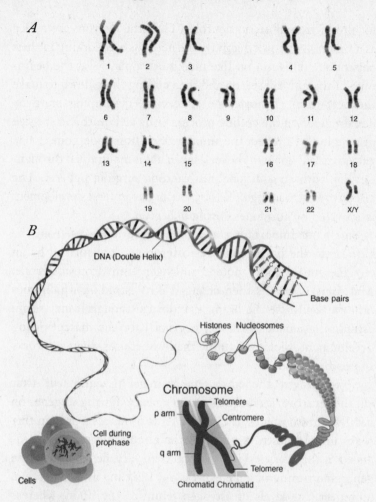

5.1 *Human chromosomes. (A) Humans have 23 pairs of chromosomes in every cell except for egg and sperm cells, which carry only one strand of each of the 23. The male sex chromosomes (XY) are shown here but not the female sex chromosomes (XX). (B) The relationship between chromosomes, DNA, and the base pairs that link the two strands of DNA. Genes are segments of DNA on the chromosomes. Only about 1.5 percent of the genes on these chromosomes (about 20,000) code for proteins, while other noncoding segments are involved in regulating the genes that do code for proteins and in other functions not yet well characterized.* Courtesy: National Human Genome Research Institute.

for production of excess neurons. Then the neurons are culled as a function of how effectively the neurons are working in their networks.[20] This is a bit like hiring extra players at the beginning of the football season and then culling them down to make the best team. Testosterone affects the developing brain by downregulating the culling process—the cell death. So the size of a region will reflect the amount of culling undergone.[21] The pattern is laid down in the fetal brain but is maintained throughout life, with an additional testosterone surge in puberty. The circuitry that was organized by hormones in fetal development is activated by hormones during puberty.

Small but important correction: Once it passes from the blood into the brain, some testosterone is transformed by an enzyme into a more potent androgen, dihydrotestosterone. And some of *that* is then changed into estradiol, which goes on to masculinize the brain. Paradoxical though it may seem, estradiol, a female hormone, is crucial for the masculinizing development. Biology is funny that way. It uses whatever works to get the job done.

So why doesn't the female fetal brain get masculinized? After all, the fetal ovaries make estrogen aplenty. Owing to genes on her XX chromosomes, the female fetus produces a protein that soaks up and destroys much of the circulating estrogen produced by the ovaries. This operation can vary, depending on the genes and timing of gene expression. Puzzling as it may seem, moreover, *low* levels of estrogen feminize the brain, whereas *high* levels masculinize it. Progesterone in the brain also helps to feminize the hypothalamus.

Couldn't I have left out that correction about estradiol? Possibly, but it conveys all by itself the fact that surprising mechanisms are in brains and bodies because we are evolved animals, not designed from scratch by a team of engineers. It also conveys a small irony. Ultimately, it is a female hormone that plays a big role in masculinizing the male brain.

Where do these testosterone-sensitive changes happen? Mostly in small regions of the hypothalamus, an evolutionarily ancient structure. Some hypothalamic nuclei regulate thirst and drinking behavior, others regulate hunger and eating behavior. As we saw in Chapter 4, some hypothalamic regions are important for parental care in mammals. Yet other regions regulate sexual behavior. These hypothalamic regions are important in the male's sexual behavior—his interest in females and his capacity to mount and penetrate the female.

The size of these cell groups in the hypothalamus is not by itself the cause of male versus female differences in sexual behavior and sexual feelings. What really makes the difference is the pattern of connections among the neurons within each group and the connections between one group and other networks in the brain. The size is just one *structural* difference that we can currently detect. Because the size correlates with stereotypical male sexual behavior, that tells us what to study next to begin to uncover the broader causal details.

In a slightly different part of the hypothalamus—an area shown in rats to be important for ovulation—the neurochemical *dopamine* inhibits cell death. In the female brain, this area expands—not prenatally, however, but during puberty. In males, the cells in this area are not only fewer in number but also different biochemically, making and releasing homegrown opioids. If by chance a female had opioid neurons in this area, they would inhibit ovulation. The cells in this area project into the nearby pituitary gland, and this is the crucial communication channel between brain and ovaries. When these neurons are active, the pituitary releases hormones that stimulate the ovaries to produce estrogen. The female cycle of release of eggs from the ovaries will then begin.

Within the male (XY) and the female (XX) populations, there is individual variation in the number of neurons in the sex-sensitive hypothalamic nuclei; that is, some men will have about the same number of neurons as some women. Averages are only averages, not invariant principles.

Another significant difference between male and female brains may help explain why females are somewhat more fearful and cautious than males. This holds not just for humans, but for other mammals as well. And I mean, of course, *on average.* So some particular female hyena may be less fearful than some particular male hyena. In female brains, the hypothalamus (the ventral medial area) is more densely connected to another subcortical structure, the amygdala. The amygdala is important for generating the fear response and for learning what is to be feared. From an evolutionary point of view, the female mammal's role in pregnancy, birth, and child rearing means that she is vastly more vulnerable than the males. She needs to be a little more cautious. To be sure, the behavioral significance of the amygdala difference is not yet thoroughly understood, and many environmental factors likely go into a particular person's profile of risk aversion and fear. Nevertheless, the finding may well lead us to a fuller understanding of these average differences in behavior.

Intricate also are the interactions between a whole orchestra of potent neurochemicals that can affect mood, personality, and temperament. They can affect risk taking, aggression, trust; whether you are shy or outgoing, easygoing or prickly. Levels vary across individuals and also within a person over time. They include such neurochemicals as serotonin, vasopressin, oxytocin, stress hormones, and somatostatin. What else is in the orchestra? Well, there are interactions involving other parts of the body, including the pituitary gland, thyroid gland, and adrenal glands. And then there is the immune system, which interacts with all these things, and the brain.

Finally, the masculinizing of the gonads (making testes, penis, and prostate) occurs before the masculinizing of the brain. Owing to variability in the pathways controlled by genes and the interactions among the items in the suite of neurochemicals, sometimes the masculinizing of the brain does not follow the typical path and may be incomplete in various ways. You could have male genitalia and a female brain.

Rats have been the model animals used in figuring out the fundamentals of the gonad-brain story. But you are not just a big rat. So, is the human hypothalamus like the rat hypothalamus? Pretty much. The basic anatomical differences in hypothalamic structures of male and female lab rats are also seen in humans. Nevertheless, human sexual and social behavior is vastly more complex. Your very large brain, especially your huge prefrontal cortex (see Figure 4.2), means that your flexibility in navigating your social world and your capacity to control sexual behavior is far more rich and varied and involves much more learning than in the case of a rat. In large-brained mammals, the interaction between gene expression and learning-based changes in the brain becomes a dense thicket of complexity.

This chapter is focused on aggression, and given the statistics on male aggression, that means we need to consider what makes a male brain male. But that is not enough. To understand the masculinization of the fetal brain at a deeper level, it is useful to see that nature does not always follow the beaten path. Reflecting on these other paths allows us a broader understanding of how hormones can affect our very nature. Then we shall return to the matter of male aggressive behavior.

DIVERGING PATHS IN
HUMAN SEXUAL DEVELOPMENT

THE BASIC account of a typical brain-hormone interaction for XX and XY fetuses has been outlined. But not all cases conform to the prototype. Variability is always a part of biology. For example, unusual chromosome arrangements can occur. Rarely, an egg or a sperm might actually carry more than one chromosome, so the conceptus ends up with more than just a pair of sex chromosomes. About 1 in 650 males are born with XXY, a condition known as Klinefelter's syndrome (as I mentioned in Chapter 1, my brother has this condition). The outcome can be quite variable, but basically what happens is that the testosterone supply dependent on the Y chromosome gets swamped by the estrogen production linked to the two X chromosomes. This affects gonad development, musculature, and fertility. There are cognitive costs, too, mainly having to do with the role of the prefrontal cortex in impulse control and the capacity to delay gratification.

Other chromosomal variations are also seen: XYY occurs in about 1 in 1,000 male births. It frequently goes unnoticed because there are no consistent symptoms. At one time it was claimed, mainly on a priori grounds, that XYY persons are especially aggressive, but this turns out not to be correct. XXYY, which is much more rare (about 1 in 20,000 male births), has many deleterious effects. This condition is linked to seizures, autism, and developmental delays in intellectual functions. In about 1 in 5,000 cases, a fetus may have only a single chromosome—an X—a condition known as Turner's syndrome. The damaging effects are very broad, including short stature, low-set ears, heart defects, nonworking ovaries, and learning deficits. If a conceptus has only a single chromosome—a Y—it probably fails to implant in the uterus and never develops.

So just at the level of the chromosomes, we see variability that belies the idea that we are all either XX or XY. What about variability in the genes that leads to variability in brain development? Various factors, both genetic and environmental, can deflect the intricate development of a body and its brain from its typical course.

Consider an XY fetus. For the androgens (testosterone and dihydrotestosterone) to do their work in its brain, they must bind to special receptors tailored specially for androgens. The androgens fit into the receptors like a key into a lock. The receptors are actually proteins, made by genes. Even small variations in the gene (*SRY*) involved in making androgen receptors can lead to a hitch. And small variations in that gene are not uncommon. In some genetic variants, the receptor lacks the right shape to allow the androgens to bind to it. This prevents the process of masculinizing the gonads and the brain. In other genetic variants, no receptor proteins are produced at all, so the androgen has nothing to bind to. In these conditions, the androgens cannot masculinize the brain or the body, despite the XY genetic makeup. In consequence, the baby, though a genetic male, will probably have a small vagina and will be believed to be female when born. This baby will grow breasts at puberty, though she/he will not menstruate and has no ovaries. This is sometimes described in the following way: the person is genetically a male, but bodily (phenotypically) a female. These individuals usually lead quite normal lives and may be sexually attracted to men or to women or in some cases to both.

If an XX fetus is exposed to high testosterone levels in the womb, her gonads at birth may be rather ambiguous, with a large clitoris or a small penis. This condition is known as *congenital adrenal hyperplasia*, or CAH. This usually results from a genetic abnormality that causes the adrenal glands to produce extra androgens. As a child, she may be more likely to engage in rough-and-tumble play and to eschew more typical girl games such as "playing house." When at puberty there is a surge of testosterone, she may develop a normal penis, testicles will descend, and her musculature may become more masculine. Though raised as a girl, persons with this history tend to live as heterosexual males. A male XY fetus may also carry the genetic defect, but in that case, the extra androgens are consistent with the male body and brain, and the condition may go unrecognized.

Some of these discoveries begin to explain things about

gender identity that otherwise strike us as puzzling. A person who is genetically XY with male gonads and a typical male body may not feel at all right as a man. So far as gender is concerned, he feels completely female. Conforming to a male role model may cause acute misery and dissonance, sometimes ending in suicide. Conversely, a woman who is genetically XX may feel a powerful conviction that she is psychologically, and in her real nature, a man. Sometimes this disconnect is characterized as a female trapped in a man's body or a male trapped in a female's body. Statistically, male-to-female transsexuals are about 2.6 times as common as female-to-male transsexuals.[22]

Is this conviction on the part of the person not just imagination run amok? In most cases, the matter is purely biological. Once we know something about the many factors, genetic and otherwise, that can alter the degree to which a brain is masculinized, it is a little easier to grasp a biological explanation for how a person might feel a disconnect between his or her gonads and his or her gender identity.

For example, in fetal development, the cells that make GnRH (gonadotropin-releasing hormone) may not have had the normal migration into the hypothalamus. If this happens, the typical masculinization of the brain cannot occur. For some individuals, the explanation of the chromosome-phenotype disconnect (XY but has a female gender identity) probably lies with GnRH, despite the existence of circulating testosterone in the blood. It is noteworthy that the data indicate that most male-to-female transsexuals do have normal levels of circulating testosterone, and most female-to-male transsexuals do have normal levels of circulating estrogen. This means that it is not the levels of these hormones in the blood that explains their predicament. Rather, we need to look at the brain itself.

Examining brains at autopsy is currently the only way to test in humans whether an explanation for a behavioral variant in terms of sexually dimorphic brain areas is on the right track.

Here is some evidence that it is. First, there is a subcortical area close to the thalamus called (sorry about this) the *bed nucleus of the stria terminalis* (BNST). The BNST is normally about twice as large in males as in females. What about male-to-female transsexuals? Because the data can be obtained only at autopsy, they are limited. Nevertheless, in the male-to-female transsexual brains examined so far, the number of cells in the BNST is small. The number looks closer to the standard female number than to the standard male number. For the single case coming to autopsy of a female-to-male transsexual, the BNST looks like that of a typical male.

What are the causal origins of this mismatch of gonads and brain? The answers are still pending, but in addition to the many ways in which lots and lots of genes could be implicated, various drugs taken by the mother during pregnancy are possible factors. What drugs? Among others, nicotine, phenobarbitol, and amphetamines.

Like many others of my generation, I first learned about transsexuality when the British journalist and travel writer James/Jan Morris was interviewed on the BBC in 1975. A highly gifted and clearheaded writer with a wonderful sense of humor, Morris discussed her long struggle with the dilemma of feeling undeniably like a female from the age of 5 and yet possessing a male body and presenting to the world as a male. In Morris's forthright book on the subject, *Conundrum*, she provides one of the deepest and most revealing narratives on what it is like to be unable to enjoy that calm sense of being at home in your own skin, of being one with yourself. Morris had married a woman to whom he was deeply attached and had five children. By all accounts, he was a wonderful and devoted father. But as the years went by, he became ever more miserable until in his 50s, and with the blessing of his wife, he underwent the long and difficult process of a sex change. Here are Morris's heartfelt words describing the change:

Now when I looked down at myself I no longer seemed a hybrid or chimera: I was all of a piece, as proportioned once again, as I had been so exuberantly on Everest long before. Then I had felt lean and muscular; now I felt above all, deliciously *clean*. The protuberances I had grown increasingly to detest had been scoured from me. I was made, by my own lights, normal.[23]

Gender identity is one thing, sexual orientation another, and the vast majority of homosexuals have no issue with gender identity at all. They just happen to be attracted to members of their own sex. This tends to be true also of cross-dressing males, who are fully content in the male gender identity but who enjoy dressing up in women's clothing. How is sexual orientation related to the brain? There are undoubtedly many causal pathways that can lead to homosexuality or bisexuality, most involving the hypothalamus in one way or another. Sometimes sexual orientation can be affected by the chromosomal and genetic variations discussed earlier. The hypothalamic changes are likely to be quite different from those found in someone who is transsexual or transgendered.

SEXUAL ATTRACTION AND ITS BIOLOGY

HOMOSEXUALITY IS one area where understanding even a little about the brain and the biological basis of a behavior has had a huge social impact. Although there are still those who maintain that homosexuality is a life choice, over the last 30 years, the credibility of this idea has largely crumbled. One significant reason was the discovery by Simon LeVay in 1991 concerning a small region of the hypothalamus (see again Figure 4.4). In postmortem comparisons of brains, LeVay found that a small hypothalamic region in the brains of male homosexuals was anatomically different from that of male heterosexuals; the region was smaller, more like that seen in the female brain. In and of itself, this discovery could not point to this brain area as the source of sexual orientation, nor could it claim with any certainty that gay men were born gay. And LeVay, tough-minded scientist that he is, was entirely clear on that point. Nevertheless, given what else was known from basic research about the role of the hypothalamus in sexual behavior, and given what else was known about fetal brain development, it seemed a good guess that being gay was simply the way some people are. More recent research strongly supports this idea.

To many people, the simple fact revealed in LeVay's anatomical discovery diminished the authority of theological arguments that homosexuality is a life choice and a direct route to damnation. Although it does not follow from the data themselves that sexual orientation is basically a biological feature, LeVay's discovery was a powerful element in ushering in a change in the culture of attitudes toward homosexuality, especially in the young, but not only in them. Antisodomy laws changed, fathers openly embraced their gay sons, and the hypocrisy of homophobic theologians and priests was exposed when they themselves were "outed." To be sure, these changes did not happen overnight, and well-organized campaigns, such as that of Harvey Milk in San Francisco, were extraordinarily important as well, but the discoveries about the brain grounded the change in attitudes in a special way.

There appears to be very little evidence that environment has much to do with sexual orientation. It is not "catching," so far as anyone can tell. Children of homosexual parents are no more likely to be homosexual themselves than children of heterosexual parents. This implies that somewhere in the thicket of causality involving genes, hormones, neurochemicals, and fetal brain development, certain of the hypothalamic cell groups are masculinized or not, to some degree or other, with the result that sexual orientation is largely established before birth. This is *biology*, not a choice made in adulthood. No 5-year-old child makes a life choice to be gay.

The main point of this section has been to discuss what is known about differences between male and female brains, especially as they pertain to testosterone. This main point benefits from being set in a broader context of the impressive variability in the biology of masculinizing or feminizing a human fetal brain. The simple idea that we are all either just like Adam or just like Eve is belied by the biological subtleties of orchestrating genes, hormones, enzymes, and receptors. One of the truly brilliant achievements of neuroscience in the last 40 years has been to make remarkable progress in explaining the link between brains and sex.

TESTOSTERONE AND AGGRESSION

UNLIKE VOLUNTARY cooperation in social birds and mammals, the roots of aggression reach deep, deep into our biological past. Even crayfish and fruit flies show aggressive behavior. In mammals, aggressive behavior depends on many neurobiological and hormonal elements. Testosterone is one element, albeit an important element, among many others. Not only are there other elements, but there are big interactive effects. Tracking interactions means tracking a dynamic system, and that quickly gets very complicated. It is a little like tracking the movement and development of a tornado. The distribution and degree of high and low pressures, warm and cold fronts, determines whether and where a tornado forms. Once it forms, the tornado itself has an effect on those very factors, which in turn continue to affect the tornado, and so back and forth. Despite the interactive nature of elements regulating aggressive behavior, some general features of the phenomenon emerge fairly clearly.

Here is one important and consistent finding. The *balance*

between testosterone and stress hormones is a strong predictor of the profile of aggression in a particular male. A disposition to be especially aggressive seems to be associated with the *imbalance* between testosterone and stress hormones. It is not just the level of testosterone per se that predisposes to aggression; it is the ratio between stress hormones (mainly cortisol) and testosterone.[24]

More precisely, male animals with high levels of testosterone and low levels of cortisol display more aggression than those with high levels of testosterone *and* high levels of cortisol. These high-high males are more vigilant and calculating. The high-low males, by contrast, are less responsive to harmful consequences that may ensue. Roughly speaking, those with higher levels of cortisol are more likely to evaluate consequences and to appreciate the nature of the risks. When high levels of testosterone balance with high levels of cortisol, the male is likely to be courageous but not reckless. Consistent with the hypothesis that the balance between testosterone and cortisol matters enormously, when levels of testosterone (but *not* cortisol) were experimentally raised in human males, fear levels decreased, even though this was not consciously recognized by the subjects themselves.[25]

Those with low levels of testosterone and high levels of cortisol are more apt to be fearful and avoid combative encounters altogether. They are particularly sensitive to risk and possible damage. They are on the reserved end of the aggression spectrum.

What controls the balance of testosterone and cortisol? Many things, including genes, other hormones, neuromodulators, age, and environment. For example, in male chimpanzees, testosterone is released when the females go into estrus.

The account is complex also because stress hormone levels can vary as a function of a host of factors. Some factors are internal, such as a genetic variant, while some are external, associated with local social practices. For example, suppose a male happens to have a very large and muscular build. In certain social conditions, he may be less stressed than a slighter male because other males will fear provoking him. In these social conditions, such individuals are treated with respect and are seldom provoked, so they come to be known as gentle giants.

If, however, more macho social customs prevail, the very same man may be tested by others precisely because of his build. In these "prove yourself" conditions, the man's stress levels may be higher. He is often provoked, so he is regularly alert for provocation and he responds accordingly. He may seem more aggressive than gentle in the stressful environment. So add the external social conventions as a third dimension.

Aggression has a fourth dimension: the neuromodulator *serotonin*. To a first approximation, levels of serotonin affect whether aggression is impulsive (low serotonin) or planned (high serotonin). Emotional responses are in play here, too, since impulsive actions tend to be motivated by very strong emotions, such as rage, whereas planned actions seem to involve more emotional control—resolve rather than rage, cagily waiting for the right moment to strike rather than heedlessly slashing away. In high-serotonin conditions, assuming that the man has a standard balance of testosterone and cortisol, then anger and fear are less prominent than intense vigilance.

Consider a male with this combination: high testosterone, low cortisol, low serotonin. This man may be especially problematic; he may be fearless and quick to anger, *and* the anger may be poorly controlled.[26] Change his ratios so that the serotonin levels become high, and his aggressive impulses will be under more control. Now consider a different male with that same profile—high testosterone, low cortisol, and high serotonin. Suppose, in addition, he rarely feels *empathy*. For the sake of argument, let us suppose that he happens to have low oxytocin levels, influencing his low empathy response. This man may be predisposed to psychopathy in his behavior. He can be aggressive, but he plans carefully and feels little anxiety during his preparations. Nor, in the aftermath, does he feel remorse for the injury inflicted. *American Psycho* portrayed such a man.

We are not yet done. Here is another, and completely unexpected, dimension (we are up to five) to aggression: a gas, *nitric oxide*. It is released from neurons and dampens aggression. It is made by an enzyme, *nitric oxide synthase*. Male animals that happen to lack the gene for making that enzyme are particularly aggressive compared with those who have the gene and hence have normal amounts of nitric oxide. In contrast to males, female animals lacking that gene are not hyperaggressive. There is some evidence for an interaction between an important neurotransmitter in the reward system, dopamine, and nitric oxide. A balance of the two is associated with control over aggression, but exactly how nitric oxide interacts with elements such as testosterone or serotonin is not understood.

There is still more. For males, defense of offspring is testosterone mediated. Vasopressin is also essential for expression of aggressive behavior against those who threaten mates and offspring. Interestingly, even with their very low testosterone levels, castrated male mice will attack nest intruders if vasopressin is administered. As discussed earlier, vasopressin is an ancient hormone that, among other things, is associated with affiliative behavior in mammals and is believed to be more abundant in males than in females. In the brain, vasopressin operates mainly on subcortical neurons, especially those associated with fear and anger. As we have come to expect, there are interactive effects with other elements in the portfolio.

To have an effect on neurons, vasopressin has to bind to receptors on the neurons and fit into those receptors, key in lock. No receptors, no effect. Vasopressin receptor density in the amygdala, a complex subcortical structure, affects the nature of the responses to fear and anger and hence affects aggressive behavior. Moreover, vasopressin receptor density itself depends on the presence of androgens, including testosterone. Additionally, androgens stimulate the genes that express vasopressin itself. Given the right balance in the hormonal and receptor orchestra, the appropriate behavior will appear at the appropriate time. The right music flows forth. But there is lots of room for dissonance and disharmony.

For females, the neural basis for defense of offspring turns out to be somewhat different from that of males. Greatly simplified, here is the story. In females, progesterone inhibits aggression. Progesterone is produced in the ovaries, the adrenals, and, during pregnancy, in the placenta. Immediately after giving birth, the mother's progesterone levels fall, but her oxytocin and vasopressin levels remain high. Oxytocin is essential to the expression of aggression, and especially in high-anxiety females, higher vasopressin levels correlate with strong aggressiveness. So much for labeling oxytocin "the cuddle hormone."[27] Biology is so much more complicated than that. The balance between progesterone and oxytocin is crucial if the mother is to be ferocious in defense of her brood.[28]

The male and female brains differ also in the density of projections from the prefrontal cortex to subcortical areas involved in aggression, such as the amygdala and the BNST. In addition, the male brain has a higher density of receptors for androgens (male sex hormones) in the amygdala and the BNST than is seen in females. Bear in mind, too, that within a population, there can be a lot of natural variability in levels of hormones, density of receptors, sensitivity to environmental stimuli, gene-environment interaction, and so forth.

Does the circuitry for aggressive behavior in predation, defense, and sexual competition overlap? Multiuse wiring is probable. Circuitry in other areas appears to be multiuse, so we should not be surprised if it is here, too. If so, a consequence is that in mammals, aggression can be sloppy. It is not always strictly suitable to the stimulus or situation.

The summary point of this section is that yes, testosterone is a major factor in male aggressive behavior, but high levels of testosterone alone do not predict that the male will be especially aggressive. Aggression is a multidimensional motivational state.

CONTROLLING AND HARNESSING AGGRESSION

AS WE will discuss at length in Chapter 7, all mammals have connections between the prefrontal cortex and the subcortical structures to manage self-control. Here we will briefly consider self-control with respect to aggression in social mammals.

During evolution of the mammalian brain, the efficient and reliable circuitry already in place in reptiles for life-sustaining behavior was expanded and modified. Basic life-sustaining circuitry that worked well in reptiles was not dismantled; pain and pleasure, pivotal in learning mechanisms, were not demolished in favor of a wholly new design. Instead, they were modified and refitted.[29] In mammals and birds, subcortical structures for aggression and defense, for fighting and fleeing, for drive and motivations (the "limbic brain") are powerful and effective in maintaining our lives. The cortex is useless without them.[30]

Pathways between the prefrontal cortex and the hypothalamus add flexibility and greater intelligence to emotional responses; they add foresight and creativity. They allow our behavior to be more considered and intelligent than instinctual.

The prefrontal cortex is richly connected to reward structures as well as to motivational and emotional circuitry. The connection to reward systems allows for shaping of the behavioral expression of emotions and drives and even for habitual suppression of certain kinds of behavior, such as aggression within the group. Social mammals learn what forms of kin competition are acceptable and what are not, and play in the young is a crucial part of that learning. Under the power of disapproval, the reward system restricts the occurrence of certain behaviors, such as aggression, to very specific conditions, such as the need for self-defense. Part of what is learned is how to shift attention elsewhere and how to damp-down the very feeling of anger. Social habits learned early are very deep in the brain's system, and they do not change easily.[31]

Hostility between groups, as well as within groups, is subject to local conventions. If you are an Inuit living in the far north, you will have grown up in a community that discourages aggressive behavior except under very specific conditions, usually involving territorial transgression or certain highly structured games.[32] The anthropologist Franz Boas reported in 1888 that it was not uncommon for someone in an Intuit camp to kill a hunter who had strayed into the group's hunting land. Nevertheless, the Inuit did not, so far as Boas could determine, engage in group warfare against each other.[33] They did have summer trade meetings, where many groups would congregate, exchanging tools and allowing courtship among the adolescents.

The Yanomamo in Brazil, by contrast, tended to encourage aggressive behavior among children, teaching them combat skills (at least during recent times when anthropologists studied them).[34] Under pressure of population growth, the adult males frequently engaged in raids against other groups. This seems to have been true also of the Haida, who lived on Haida Gwaii (formerly the Queen Charlotte Islands), off the north coast of British Columbia. In this they contrasted with the neighboring Tlingit and Salish tribes, who were raided, but usually not themselves raiders. (See also Chapter 6 for more discussion of these differences.)

We live in a matrix of social practices, practices that shape our expectations, our beliefs, our emotions, and our behavior—even our gut reactions. Our personalities and temperaments are bent and formed within the scaffolding of social reality. The matrix gives us status and strength, and, above all, predictability. The matrix of social conventions is both a boon and a bind—sort of the way a sailboat is. You need it to move through the waters, but you have to play by its rules.

Different sailboat configurations serve the sailor in different ways, and some of these configurations were not always obvious. The keel, so critical to sailing, was first invented by the Vikings around 800 CE, giving them a huge advantage in sea travel, not to mention conquering and pillaging.[35] Likewise, social institutions, such as having an independent police force paid by taxes on the citizenry or allowing women to vote, were not obvious, at least not until they were put in place. Once tried with success, social institutions tend to stabilize, mature, and spread. They become second nature. Their justness seems so transparently, unmistakably self-evident. We make up myths to depict them as universal among all civilized humans, as having been in force since the dawn of *Homo sapiens*, or perhaps as having been handed down by a supernatural being.

A MAJOR target of exploration of this chapter has been the association of aggression and hate with pleasure. Oddly, there is very little research on the neurobiology of this association and not much psychological research either. At several points in thinking about the link, I have wondered whether I am just plain wrong in perceiving that the link exists. I suspect that I am not wrong. I expect that hostility does not always involve pleasure, but in some conditions, particularly when groups fight groups, the two seem closely linked. Evidently, also important is the hormonal balance and the receptor density and distribution for the various hormones. To a first approximation, human males and females display differences in aggressive behavior that are linked to male and female hormones, though these behavioral dispositions can be modulated by the cultural matrix.

At the same time, aggressive impulses in all mammals are subject to self-control. But before taking a closer look at how the brain regulates self-control, I want to explore a more basic question in the next chapter: Do our genes dispose us to warfare against other humans?

Chapter 6

Such a Lovely War[1]

"The rush of battle is a potent and often lethal
addiction, for war is a drug, one I ingested for
many years."

—CHRIS HEDGES[2]

IS GENOCIDE IN OUR GENES?

HERE IS WHY this is a tricky question. For *anything* you do,
you must have the *capacity* to do that thing, and hence you
must have the wiring for that capacity. Otherwise, lacking the
capacity, you could not perform that action. So if humans can,
and regularly do, kill other humans, they must have the capacity for it and hence the wiring for it. Fine. But what about the
hypothesis that killing other humans was *selected* for in our evolutionary history, and that is why we do it? It says that the wiring
is there because there are genes specifically for the wiring that
supports killing other humans, and those genes were selected

for in human brain evolution. For *that* kind of a claim, evidence goes begging.

To claim that genocide is in our genes on the grounds that humans do commit genocide would be like saying that we have genes for reading and writing because humans do read and write. This latter we know to be wrong. Writing and reading were invented a mere 5,000 years ago and were made possible by other, more general capacities that we have, such as fine motor control in our hands and the capacity for detailed pattern recognition. Writing and reading were cultural inventions that spread like wildfire. Writing and reading *seem* as fundamental as anything we do. But they were a cultural invention. Moreover, it took about 250,000 years before humans invented writing and reading, so they could not have been *obvious* cultural inventions. Indeed, some aboriginals living in the Americas such as the Inuit never did invent writing, though they were remarkably inventive when it came to tools and boats.[3]

From all that we now know, human warfare was not *as such* selected for in biological evolution. It may have been, like reading, a cultural invention that exploited other capacities. Perhaps the pleasurable bloodlust recruited during the hunt of large mammals such as buffalo and bears is recruited also for the hunt of humans. In some environments, humans may have found other humans a good source of protein.[4] Depending on conditions, it may have taken little to extend killing from buffalo to outgroup humans. Certainly, our human Stone Age ancestors had to be courageous, clever, and persistent to hunt nonhuman mammals. They had to be energized for the hunt and willing to hack and club animals for food. The men, like wolves in a pack, needed to be bonded to each other to succeed in a dangerous and difficult task.[5]

Whether the hacking and clubbing typical of the hunt were simply extended to other humans as human populations grew and the profit of raids and plunder became more evident is uncertain. What is clear is that by the time human events began to be recorded some 5,000 years or so after the advent of agriculture, raiding and conquering, along with the taking of slaves, was well under way.[6] Once the practice of war got a foothold, it seems to have had stubborn staying power—just like writing, body decorating, and wearing shoes.

Could genes explain the differences in attitude to raiding and war between the Inuit and the Yanomamo? Conceivably, but other explanations need to be put on the table before rushing to assume that differences in genes explain the differences in the customs regarding warfare. Ecological conditions make a major difference to whether intergroup violence takes hold as a practice. Consider that the winter camps of Intuit are small and separated by long distances. Sneaking up on an igloo is next to impossible, given the open terrain and the early warning provided by the sled dogs. So the Arctic terrain is inhospitable to war. The Inuit, though not keen on warfare, were certainly tough and cunning predators. Had it been profitable, they might have engaged in war. For the Inuit, hunting seals was virtually always more profitable than a raid, which is unpredictable, risky to life and limb, and unlikely to yield anything of value (though the Inuit did raid Cree camps to the south for women, likely a strategy to avoid the disasters of consanguinity).

In the Arctic summer, the camps were larger, but again the distances between any two camps made raiding prohibitive. Launching a sneak attack in the treeless Arctic is even more futile in summer, given the length of the midsummer photoperiod: 24 hours of sunlight; no darkness for cover. Hunting, fishing, and caching the meat makes more economic sense than risking an energy expenditure on a raid with little in the way of expected payoff. Remember that for those living above the Arctic Circle, starvation was a constant threat and a frequent reality. Energy expenditure was always a factor to calculate in a plan. In a climate so harsh and unforgiving, stupidity and miscalculation were well understood to lead to catastrophe.

The Yanomamo, by contrast, had developed agriculture, which provides a fairly reliable food source. Since the women do most of the fieldwork, the men are rather more free to plot and practice and prepare for raids. And the Amazon jungle, unlike the tundra, provides cover for sneaking around. Inuit and Yanomamo children grow up in different cultures, and the reward systems of young brains are tuned in very different ways, which affects their acquisition of social practices and norms and the development of habits of cortical control. This also yields, on average, different intuitions about how to behave socially. It is probable, but not certain, that what is important are differences pertaining to resource pressures as well as to the costs and benefits of carrying out a raid.[7] On the other hand, it could be that warfaring dispositions *were* selected for but are suppressed in the nonwarring cultures.[8] This is what the "genes for warfare" advocates will want to suggest, but they will need evidence if they wish to make it a statement of fact. It will be up to the geneticists to identify those genes, if they exist. Anthropological speculations about "genes for" do not constitute genetic evidence.

In sum, anthropological reports show that not all hunter-gatherer groups made war on other groups. Instead, it looks like there may have been quite a lot of variability across cultures. In considering the hypothesis that genes for warfare were selected, one view is that very early humans, living about 250,000 years ago, were likely quite sparsely spread around Africa and may not have encountered each other very frequently. So as long as resources were adequate or new territory was available, wars between groups might have been too costly to have been worthwhile. There is evidence of trade among very early humans, and it is not out of the question that amicable trade and exchange of females made more sense than hostile clashes. Some evidence of violent interactions comes from Stone Age skulls bearing marks indicating violent death, but whether this was an unusual instance or a common occurrence is really not known at this point. Needless to say, evidence regarding the social life of early humans is rather hard to come by.

What can geneticists tell us about genes for warfare? First, geneticists are leery of framing the question that way. For one thing, genes are bits of DNA that code for amino acids. The DNA code is transcribed to RNA, which strings together amino acids to make a protein; genes do not directly control behavior. For another, there is a long and complicated causal route between making proteins and building neural circuits out of neurons. This requires interactions between networks of genes, between genes and the developing brain, and between brain structures and the environment. So the causal route between a gene and behavior is not only long, but the very opposite of direct, the more so when that behavior is not a reflex, like the eyeblink reflex, but a cognitive decision, such as "let's massacre the folks in the next village."

As we saw in Chapter 5, aggression involves many different elements—a palette of hormones, neuromodulators, receptors, enzymes, not to mention the circuitry to make it all go and the right arrangement of receptors and hormones during fetal development. These all depend on gene expression, and that expression in turn is regulated by other genes that may be regulated by yet other genes sensitive to features of the environment. Even if we know that a neuromodulator such as serotonin plays a role in aggressive dispositions, the road to the conclusion that the gene for serotonin is a gene for warfare is long and tortured. My claim requires some backfill to be made good. The backfill is best provided by what I like to call the parable of aggression in the fruit fly.

In fruit flies and mice, a connection between a neuromodulator, serotonin, and aggression has been observed.[9] Experimentally elevating levels of serotonin using drugs or genetic techniques

increases aggression in the fruit fly; genetically silencing serotonin circuits decreases aggression. These results are, moreover,
consistent with experiments on the mouse, suggesting conservation of mechanisms for aggression through evolutionary change.
Given these data, you might predict that the gene that expresses
serotonin should be known as the "aggression gene." Not so fast.
Let's test the idea.[10]

Geneticists Herman Dierick and Ralph Greenspan[11]
selectively bred aggressive fruit flies for even more aggressive
behavior. After 21 generations, the male fruit flies were 30 times
more aggressive than the wild-type flies. Just like breeding dogs.
Next, they compared the gene expression profiles of the aggressive flies with that of their more docile cousins using molecular
techniques (microarray analysis). If serotonin is the "aggression
molecule" and the gene for serotonin the "aggression gene," this
experiment should reveal it to be so.

The surprising result was that no single gene was fingered as
specifically associated. Instead, small expression differences were
found in about 80 different genes.[12] What genes were they? *Not
the genes for regulating expression of serotonin.* Many of the genes
whose expression had changed were known to play a role in a
hodgepodge of phenotypic processes—cuticle formation, muscle
contraction, energy metabolism, RNA binding, DNA binding,
development of a range of structures including cytoskeleton—as
well as many genes having unknown functions. *No single gene on
its own seemed to make much difference to aggressive behavior.*[13]

How can that be, given the earlier experiments showing that
elevating serotonin levels enhances aggression? First, as already
mentioned, the relationship between genes and brain structures
does not remotely reflect a simple "gene-for" model. Genes are
part of networks, and there are interactions between elements
of the network and with the environment. This is a huge challenge for someone like Jonathan Haidt, who claims that there
are genes for being liberal or conservative.[14]

Second, bear in mind that serotonin is a very ancient molecule. It is important in a motley assortment of brain and body functions: the list includes sleep, mood, gut motility (such as stomach and intestinal contractions), bladder functions, cardiovascular function, stress responses, induction of smooth muscle proliferation in the lung during embryological development, and regulation of acute and chronic responses to low levels of oxygen (hypoxia).[15]

The point of this list is to dramatize the diversity of jobs that serotonin affects and hence the glaring unsuitability of the label *the gene for aggression*. The diversity of serotonin functions also helps explain how changing its levels can have widespread effects all over the brain and body, and these effects can cascade into other effects, which may in turn exert an influence on aggressive behavior.[16]

The main lesson of the parable of aggression in the fruit fly is that it is easy to speculate about a gene-for aggression based merely on observation of a behavior and perhaps an intervention, such as experimentally altering the level of serotonin. But unless you do the genetic tests, you have no clue whether your speculation will be validated by science.

If the relation between genes and aggression is that messy in fruit flies, how likely is it that the simple "gene-for-genocide" model applies to humans? Not even marginally likely. This is not to say that genes make no difference to aggressive behavior. They absolutely do, as the Dierick and Greenspan experiment clearly shows. But the causal relationship between a gene and the brain structures involved in aggressive behavior is a vast and elaborate network of interacting elements.[17] Moreover, some of those brain structures are responsive to the reward system, which modulates the likelihood of aggressive behavior toward other humans as a function of sensitivity to cultural norms. We saw the reward system at work in the contrast between the Inuit and Yanomamo customs regarding the expression of aggression.[18]

Back to warfare: so far as the evidence is concerned, it is conceivable that warring behavior was selected for as such, but it is vastly more probable that it is mainly an extension of other dispositions, such as the bloodlust needed in predatory behavior. The bane of evolutionary psychology as a discipline has been a regrettable tendency to announce that some identified behavior was selected for and then to cobble together a story about our Stone Age past to explain why.[19] The problem is that inventions of the imagination, fun and coherent as they may be, do not constitute actual evidence. It is quite another (and more difficult) thing to actually show that there are genes linked to the brain structures that are linked to *that* specific behavior, as opposed to some other more general behavioral capacity. Generally, evolutionary psychologists do not even try to do this.[20]

A small handful of human diseases, such as Huntington's disease, are associated with a single gene. This tight link between a single gene and a phenotype is the exception, not the rule, even in the case of diseases. Even height, commonly suspected to be linked to a single gene, is linked to many genes. There are 50 known genes associated with height, and how many other genes are involved but not yet identified is anyone's guess. The idea that any aspect of human cognitive behavior, such as genocide, is tightly caused by a single gene or two is unlikely.

What is often overlooked is the fact that many forms of behavior can be widespread even though that behavior was not selected for in the evolution of the brain. As Elizabeth Bates famously pointed out, feeding by using your hands is universal among humans. But it was probably not selected for as such. It is just a good way of getting the job done, given the body's equipment. Better than using your feet, anyhow.

In sum, it is better to acknowledge our ignorance than to make up a fetching story about what is in our genes, a story that, for all we know, may be dreadfully misleading.[21] One enduring problem with the "it's in our genes" refrain is that it is sung out as an excuse for bad behavior: "What can we do? It's in our genes!" Thus are greed, warfare, racism, and rape put on a plane of respectability.[22] The fewer myths we have about aggression and its origins, the better our chances of avoiding policy decisions that are counterproductive.

HOW INSTITUTIONAL NORMS SHAPE BEHAVIOR

FROM THE PERSPECTIVE of the brain, one major advantage of cultural norms is that they reduce uncertainty. You know what you have to do and what I am likely to do. We all know our jobs. You do not have to stop and figure it out each time. Norms allow prediction of the social future to be vastly easier. Novel situations are demanding in terms of attention, vigilance, and the anxiety of uncertainty. The greater the degree of life-relevant novelty, the greater the energy expenditure. The death of a parent or respected national leader puts you in disarray. What happens now? A chunk of the matrix of social stability has suddenly collapsed.

Brains love predictability and are organized to learn so they can get it. They will learn practices and norms that provide social predictability. Basically, it matters little whether you have the convention of belching loudly after a host's wonderful meal or of suppressing that belch and offering verbal expressions of gratitude. What matters is that you do not have to wonder what to do and be anxious about a misstep. In Japan it is expected that you noisily slurp your noodles to indicate pleasure. Now that you know, you can settle into slurping without worrying that you might be violating a social norm, as you certainly would be if you were eating soup at high table in Oxford.

What matters is that having a convention means you do not have to put energy into figuring out whether you should do one thing or the other. Slurp in Japan, not in England. Easy. You play your expected role. You feel comfortable. It is right. You are acceptable. You can think about more important things.

The norms of institutions and social practices are not always explicit. You may unconsciously pick up patterns of behavior as "just the way things are done." You model yourself after those you consider admirable. You want to be like them. You imitate gestures, contours of speech, and even a whole suite of behaviors. You imitate fashion in clothes and haircuts. As part of the group, you want approval and not exclusion, so challenging a norm will not come easily. Conformity to local norms feels comfortable. Speaking out or resisting is edgy. It sends the levels of stress hormones climbing.

You know how to act if you must be a witness in a court case. You know that you are expected to tell the truth, and you know about the penalty for perjury. You know what to do if you are a member of the jury or if you are an attorney for the accused. Institutions, such as the criminal justice system, may not be perfect. Still, if you believe they work reasonably well, you put your trust in them. You behave pretty much as the institutional structure guides you to behave. It is therefore shocking when you learn that a judge was bribed or a prosecuting attorney faked the evidence.[23]

A different set of norms applies when you contract with a plumber to install a new shower or with a physician to lance a boil or with a teacher to learn cellular biology or with a minister to conduct a wedding. Institutions such as colleges, churches, and businesses embody cultural norms that structure our behavior as we engage with them. You slip in and out of roles very easily, conforming to this or that norm.[24]

The television series *Mad Men* shows us highly conforming role-fitting in an advertising agency. The men largely behave in essentially the same manner: smoke constantly, drink constantly, bed the female staff, and ruthlessly do whatever it takes to come out on top of a deal. Small variations exist, but they pale next to the conformity. Most of the females are secretaries, fawning over the men, mostly acquiescing in the unquestioned assumption that high pay and good work is for men, and men alone.[25] In a different television series, *The Sopranos*, a different set of norms governed the lives and behavior of the men in the Mafia. Things the advertising men in *Mad Men* would find horrifying, such as "whacking" a suspected informer, the men in *The Sopranos* considered just part of a day's work.

Aristotle, along with many thinkers after him, recognized the importance of social practices and institutional norms in regulating behavior and providing the scaffolding for many interactions, thereby undergirding stability and harmony. Thus, some norm or other for conflict resolution, for preventing conflict from destabilizing the entire group, is seen in all cultures. When the institutions themselves are problematic or corrupt, people may drift, slowly and with little reflection, into roles where they do truly evil things. Or they may get fed up, band together, and try to modify the institutions.

Here is an example of what is clearly a cultural phenomenon. In geographically distant regions within the United States, different expectations and norms guide how men respond to insults. In their research, Cohen and Nisbett studied the effect of the "culture of honor" in the South, as contrasted to very different norms prevailing in, for example, Minnesota.[26] A southern man is expected to fight in response to an insult or risk losing face and social standing. A man in Minnesota, by contrast, is more likely to find such a response to an insult excessive and foolish, especially as the insult may itself be regarded as a sign of social ignorance and hence not worthy of notice or energy expenditure—unless it is, in which case, bide your time and get back at him later, when you can inflict real damage.

I am watching the movie *Inside Job*, a documentary about the financial crash of 2008. I ponder the men at Goldman Sachs who, according to their own e-mail messages, knowingly sold their trusting clients securities that they evaluated as "a load of crap." Moreover, they themselves bet against those very securities and stood to make a huge profit when their clients' securities became worthless. Totally corrupt. They did in fact make huge profits betting against their clients.

To get fees, many companies sold mortgages to people who could not possibly keep up the payments and who did not understand that low "teaser" interest rates were soon to rise to unmanageably high rates. The crash, of which the Goldman Sachs infamy was only one segment in an avalanche of comparable infamies, destroyed the financial lives of millions of people. Moreover, the documentary claims, the Wall Street big boys and the mortgage companies got away with it.

I am tormented by these thoughts: Am *I* capable of participating in such unflinching, venal destruction? Is it just luck that I have not been party to horrific fraud or some appalling massacre of innocents?[27] If I had been part of that Wall Street culture, would I have fallen into step? In the subculture of Goldman Sachs securities trading, would I have suckered clients to get a profit?

Experiments in social psychology, such as Philip Zimbardo's pioneering Stanford Prison Experiment,[28] in which male Stanford students were randomly assigned roles as prisoners or guards, are revealing. Famously, the experiment had to be terminated, as the "guards" became vicious and abusive to the "prisoners." Zimbardo's experiment showed that our long-standing attitudes and beliefs, our temperaments and personalities, are vulnerable to shifts if the situation suddenly changes and a new set of norms prevails.[29] Although norms are vulnerable to a shift as social background shifts, this is not *inevitable*, especially if we are forewarned and alerted to the change. Still, the worry is that in the new world of unregulated derivatives and unregulated default swaps, where everyone else is "doing it," it is not improbable that I would have let my scruples soften. Still, not everyone did, and some even tried to blow the whistle.

Because we depend on our social institutions for norms concerning justice, prosperity, and decency, we have to ask: What happens when those very institutions subvert decency? What happens when the brain's reward system gets tuned to social practices that are evil? What happens when clannishness becomes cruel or fraternity hazing gets so bad that the men end up killing an initiate? It is right here that human responsibility is of monumental significance. Collectively, we are responsible for our social institutions. They are rooted and structured and refurbished by us. It is right here where Socrates, sentenced to death on a trumped-up charge of corrupting the youth when in fact he merely raised questions that were embarrassing to the authorities, refused to back down. It is right here where Nelson Mandela convinced the crowds intent on revenge, after years of apartheid, to take the peaceful option.

In one of the most profound books on the moral history of the twentieth century, Jonathan Glover sees the issue this way:

> When terrible orders are given, some people resist because of their conception of who they are. But there may be no resistance when a person's self-conception is built round obedience. In the same way, if someone's self-conception is built round a tribal identity, or round some system of belief, resistance to tribal or ideological atrocities may have been subverted from within. A lot depends on how far the sense of moral identity has been narrowed to a merely tribal or ideological one.[30]

Glover suggests that one important tool in staying on the right side of the dilemma is skepticism, not believing too readily the emotional appeals in politics, economics, or anywhere else for that matter. Think. Get the facts. Read history. Get a bit of distance.

Another important tool is a lifelong determination to stand in the truth, not to let it be watered down into gauzy, cheap "truthiness." Here again, Glover goes straight to the heart of the matter:

> In an authoritarian society there is much more power to defend the small lie by massively churning out the larger one. It is like printing money to get out of a financial crisis. There is short-term gain for the authorities, who avoid being immediately found out. But the long-term cost of this propaganda inflation is that the official belief system grows ever further out of touch with reality, and so hard to support without yet more coercion and lying.[31]

———

WARFARE, so much a part of human life since historical times, can seem inevitable. Perhaps, however, it is not. Perhaps it is no more in our genes than using fire or making boats. Nevertheless, it has been remarkably appealing. For example, in the aftermath of the attack on the World Trade Center (9/11), the U.S. government met little opposition in its plans to go to war against Iraq, a country that no one had any serious reason to believe played any role in the attack. On an optimistic note, however, Steven Pinker, relying on the long view and a close analysis of the statistics of war, argues in his most recent book that modifications to institutions can reward less deadly ways of handling conflicts, slowly adjusting the way business gets done.[32] Warfare on a large scale, he suggests, can become outmoded, and perhaps it has already become less inevitable than it was a hundred years ago. I find I want to be convinced by Pinker's argument, but I am still wary. If prosperity were to plummet and food become scarce, are our institutions strong enough so that war would be considered an unacceptable option?

History shows us how little it takes to subvert institutions and to promote the big lie, as Glover calls it. The stability of our institutions can seem immune to onslaught, but in fact they are terrifyingly vulnerable. Sometimes I fear we are doomed to yield to the constant pressure of propaganda, just as, incredible as it now seems, the German people did so yield in the run-up to the Second World War. How little it takes to whip up fear, panic, and the impulse to kill, all with the best will in the world, knowing, with the enthusiasm of the righteous, how gloriously right and justified one is.

The nagging worry that is apt to surface now is this: Do we have any self-control, really, if so much about our behavior is organized by nonconscious events in our brains? If we are pulled hither and thither by hormones and enzymes and neurochemicals galore, isn't self-control just an illusion?

Self-control is something we are encouraged to develop as children, to exercise as adults, and to maintain in the face of enticing temptations and terrifying threats. What is control in neural terms? How do brains with weak self-control differ from those with strong self-control? And if self-control is not real, as the phrase "free will is an illusion"[33] suggests, then what is the point of trying to make reasonable choices—of trying to be responsible, courageous, decent, and honest?

Chapter 7

Free Will, Habits, and Self-Control

BRAIN MECHANISMS FOR SELF-CONTROL

BUTCH FANCIED himself quite the ferocious watchdog. He was a terrier who belonged to the MacFarlanes, our kindly neighbors with a grand orchard to the south. Despite their reassurances that he would not bite, I could not but notice that when Jordie MacFarlane dropped by our house, Butch eagerly picked a fight with Ferguson, our German shepherd who was three times his size. The fighting was an event that called for a water hose, turned on full blast, something I may have focused largely on Butch.

Required to deliver two dozen eggs to Mrs. MacFarlane, I was instructed by my older sister: "No matter how frightened you are, never, *ever* run from Butch. Walk forward slowly when he thunders off the porch with teeth bared, look past him, and think serenely of something you really like to do. Never show fear. *And don't drop the eggs.*" I practiced in the garden a few times, rehearsing especially not running, and trudged off. Butch did his worst, I thought of swimming in Osoyoos Lake, and my 5-year-old self outbluffed the hated Butch.

Butch mastered, more eggs were delivered past other fearful guardians, and eventually I found myself confronting worse adversaries—such as Patrick O'Donnell, who liked to bully the younger boys on the school bus. I think, *Never show fear, and hit him hard on the noggin with my lunch box.* Done. He stops. Mrs. McCormick, our bus driver, turned a blind eye and a deaf ear, bless her. Or, as a young professor, when a few male colleagues got about as much fun out of bullying female faculty as Butch got out of bullying me. *Never show fear. Stay calm and carry on. Outlast the bastards.*

In facing down Butch, I was learning lessons in self-control. I was growing the skills needed in facing assorted threats, just as generations of youngsters, humans and otherwise, have always done. Learning to exercise control is important not only concerning fears, but also concerning various temptations, impulses, emotions, and seductive choices. You learn to use imagination and rehearsal to achieve what seems so difficult. You take satisfaction when control works out well. You feel remorse when you lost control and paid a price. Habits form, so that you do not always have to work at maintaining control. Your brain's control system gets stronger.

Much of early life is about learning self-control and acquiring the habits of suppressing costly impulses and doing things we would prefer not to. In acquiring these survival habits, we are guided by parents and mentors. "Chores first, play later," "Serve guests first, yourself last," and so on. Much of later life is about exercising control and allowing judgment its space to influence decisions. All those virtues we are encouraged to develop—courage, patience, persistence, skepticism, generosity, thrift, hard work, for example—are rooted in self-control. And these virtues are entirely real, not illusory in the way that Earth's appearing to be flat is an illusion.

But what really is self-control? Plato thought about self-control in terms of a metaphor: reason is the charioteer, while the headstrong horses are emotions and the appetites. You are in the chariot. Wait: I thought I *am* the mix of reason, emotions, appetites, and decisions. What, from the perspective of the brain, could be the charioteer? How does Plato make *that* work out? Maybe neuroscience can give us a more insightful slant.

Let's go back to mammals. Unique to the mammalian brain is the *neocortex*, a six-layered mantle of neurons covering the subcortical structures (Figure 4.1). The high cost of mammalian dependency at birth is offset by the great advantages of new kinds of learning, such as imitation, trial and error, and the ability to recollect particular events and places. All kinds of learning greatly expanded in mammalian evolution. The problem of helplessness at birth is also offset by the increased capacity for control. Self-control and being smart are achievements of the neocortex and how it interweaves with the ancient subcortical structures.[1]

To the neocortex is owed the spectacular increase in mammalian intelligence and capacity for self-control, though precisely how this all works in terms of neural mechanisms is not known. In primates, the great expansion of the prefrontal cortex is probably also responsible for the enhanced capacity to predict the behavior of others, usually via attribution of goals and emotions. In humans, our capacity to understand what is going on around us involves drawing from a complex palette of mental states. "He is angry," "She has low self-esteem," "He is grouchy when he has not had enough sleep," "Uncle George is delusional; he thinks he is Jesus."

Seeing the behavior of ourselves and others as the product of mental states such as fearing and hoping and wanting helps us organize our social world. It allows us better predictions concerning what others will do, especially in response to what we do. That gives our self-control circuitry information about what to avoid and what is required to get on in the social world.[2]

In the evolution of the brain, intelligence and control grew more powerful in tandem. This makes sense. If I can accurately predict that you will beat on me if I steal from you, that knowledge does me no good unless I can act on it—that is, unless I can control my impulse to gain by stealing. By and large, children are typically well able to acquire control of these sorts of impulses, given a guiding, loving hand.

To be sure, you can expect individual variability in the capacity for self-control as you can for just about every other mental capacity, such as risk aversion or sense of humor or recall of past events. You notice that some adults find it relatively easy to be thrifty, while others are regularly hooked by slick advertising, buying junk they do not really need. Some adults are always finding the funny side of little gaffs, while others are invariably glum, even when you think the funny side is downright hilarious.

One test for differences in self-control in preschool children goes like this: "Would you like one marshmallow now or three marshmallows in five minutes?"[3] Some children are able to wait for the bigger prize, but some go for the lesser but immediate option. Remarkably, whether they choose to defer gratification or not turns out to predict fairly well how they will fare later in life in those endeavors where self-control is a factor.[4]

Of the children who can wait for the bigger reward, on average more go to college, fewer are obese, fewer are drug addicts, they get better jobs, they have fewer divorces, and in general they are healthier. Even so, it is important to realize that self-control can be improved in children, including those who find it hard to defer gratification, by teaching them to imagine themselves making the better of two choices or to imagine themselves holding back on their impulse to hit another child. Self-control and attention is also improved if the children engage in role-playing: "You be the policeman and I will be the robber"; "I will be the teacher and you will be the student."[5]

That self-control can be enhanced by these simple tactics

is an important reminder that even though the role of genes is immense, it is by no means the whole story. We are highly adaptable. Our big prefrontal cortex can be trained.

Individual variability in the capacity for self-control is seen in many animals. For example, some dogs are easier than others to train to stay on task and ignore diversions. Golden retriever puppies are selected for further training as handicap dogs on the basis of temperament and how easily they can be trained to control their impulses. Our retriever, Max, could easily be trained to ignore an approaching dog and to look straight ahead. Fergus, his littermate, never could. He would hold back for a few seconds, give in, and go jump gleefully on the approaching dog.

Rats, too, differ in the capacity for self-control. Here is the way this is shown in Trevor Robbins's lab at Cambridge University: A food pellet dispenser is set up that the rat can activate by poking his nose on a button. Right after the rat nose-pokes, one pellet is delivered down a chute. This pleases him, so he does it again. Once the rat has the hang of it, a second dispenser is added.[6] The second dispenser yields four pellets after a nose poke. Not stupid, the rats quickly learn to activate the high-paying dispenser. They get only one poke per trial, and then the rats are removed pending the next trial. So if the rat pokes at the one-pellet dispenser, that is all the reward he can get for that trial.

Here is how to separate out the rats with better self-control. In the high-paying dispenser, a time delay is introduced between the poke and the delivery of pellets—first 10 seconds, then 20 seconds, and eventually 60 seconds. Some rats can wait up to 60 seconds for the four pellets; some cannot wait and rush back to poke the one-pellet dispenser. This is a clean measure of a certain kind of self-control, namely, deferred gratification. Do the rats who fail to defer gratification feel remorse over their stupid impulsiveness? That is hard to say, since there is no good measure for a rat's remorse.

A different sort of experiment clocks the reaction time of the rat in stopping an action after initiating it but before completing it. A tone tells the rat to stop its action, and the experimenter can vary just when in the rat's movement to deliver the stop signal. Lest this all seem a bit ecologically invalid (silly), the capacity to stop a planned action is believed to allow for flexibility in changing environments, as when you are about to throw a rock and a bear crosses the presumed trajectory of the rock or you are about to pitch to the batter and the runner on first takes off for second base.

Some rats can abruptly stop even when they have moved so far along that they have begun to put their nose toward the delivery button; some cannot stop even when the tone sounds well before that. These are called stop-signal reaction time experiments and can be easily adapted for human subjects. It might be supposed that the same rats that are easily able to defer gratification will also be champs in the stop-signal reaction time experiments, but this is not entirely how things turn out. Additionally, serotonin depletion in rats decreases their ability to wait but has no effect on their reaction time in a stop-signal reaction time experiment. The results suggest that the two manifestations of self-control—deferring gratification and response inhibition—are at least somewhat dissociable. This is interesting, because it means that self-control is not one single capacity, but perhaps interwoven capacities that share a widely distributed neural substrate. Maintaining a goal in the face of distraction, also a measure of self-control, may be a different element in the control portfolio. Overlapping but somewhat distinct circuitry may be involved in goal maintenance.

If you want to see control in the wild, watch a predator stalking prey. A fox slowly, silently creeps up on the pheasant until she is close enough to spring with a good chance of success. Her pups watch and learn. When a pup first tries for himself, he is apt to spring too soon. Feeling the disappointment of losing out on food, he learns to bide his time. He learns to control his impulse to spring *now*.

Or watch a grizzly bring down a caribou.[7] As her two cubs observe on the sidelines, the sow grizzly lures a modestly lame caribou into a rocky stream with brief charges, always anticipating his fitful but dangerous countercharges and cannily edging the caribou ever closer to deep water with each of his charges at her. Finally in deep water, she lunges straight at him, lodging her body between his fearful rack of horns, and with her enormous weight twists his head until he cannot stand. He tips over into the deep water. Then she lays her body on his head, and he slowly drowns, his most powerful weapon, his back legs, now frantically kicking the air in vain.

This is a dramatic exhibition of intelligence and self-control. The grizzly first has to target a somewhat handicapped animal. Then she has to modify her actions according to changing circumstances, keeping track of the whereabouts of the cubs, grunting at them if they move in too closely. She has little chance of bringing down the caribou on land. It would be a standoff. So she methodically baits the caribou into charging at her in the water and waits for the kill until the moment is right and the water deep. This is impressive self-control.

Human hunters display similar uses of strategy, knowledge, and getting the timing just right. Once on an eco-trip with students to the Arctic, our Inuit guide smelled a musk ox herd nearby. These animals are skittish and would leave the river area, disappearing into the tundra, unless we knew how to approach. Going downwind, we then, as silently as possible, edged forward inch by inch on our bellies until we saw the magnificent creatures with their amazing long hair and huge horns.

Cooperative hunting requires both self-control and knowledge of what your job is, relative to what others are doing. Social intelligence as well as experience in teamwork is an additional factor in success. Wolves are masters in team hunting and in adjusting to opportunities, moment by moment.[8] Humans are, too. Human teamwork has the advantage that language and other symbols can be used to fine-tune arrangements, but other animals, such as wolves, use nonlinguistic cues, such as whistles and yelps.

Control in aggression is highly advantageous. An animal needs to have a sense of when to back down in a fight, when fighting is likely to provoke overwhelming retaliation by the group, when fleeing a predator might be better than fighting him, and so on. Aggression, like everything else involving survival, requires judgment. In some fashion, the control system is also the system supporting judgment. It is a system that balances risks and weighs probable gains against probable losses in both the short term and the long term.

Panic is often the wrong response to an emergency; cooler heads are apt to manage better. Disgust and avoidance can be the wrong response when a bone fragment needs to be pulled from the throat of a dog, or porcupine quills from a child's arm. If someone else is around to do the job, you might duck it, but often no one is. You have to do it and not be squeamish about it so it can be done properly.

Loyalty to a brutal and crazed leader can lead to worse horrors than defection. Pride can mean that opportunities are squandered and the fall inevitable. Getting the balance more or less right is something for which we have no rules, only experience and reflection on experience. Balance, judgment, and common sense are part of what our prefrontal structures, in combination with our basal ganglia, are good at.

In helping others, lack of control can be a disadvantage. Sometimes a rush of empathy in the face of another's misery may cause you to overreact and to create expectations that cannot be sustained. Sometimes those in need manipulate and take advantage of the kindly hearts of those whose empathy is stirred and who sacrifice easily.[9] Sometimes the need is itself a charade, staged for exploitation of those who do not look too closely. Balance and judgment again are called for.

We can see that in the evolution of the mammalian brain, intelligence and the capacity for self-control reached new levels of complexity. Predators got smarter, but prey got smarter, too; then the predators got smarter yet again.

Neuroscientists know in a general way what structures are crucial for normal self-control in its various manifestations. Even this general knowledge is fairly recent, and very little is known about exactly how the players conduct their business. Such knowledge will eventually be unearthed, but it will be a piece of the larger story of how billions of neurons work together and do so without a conductor, without a commander in chief. With those cautions nailed up, we can see that self-control depends on the connectivity patterns between neurons in a set of subareas of the prefrontal cortex (PFC) and subcortical structures, mainly the basal ganglia and nucleus accumbens (Figure 7.1). Different functions in the self-control portfolio may share cortical regions in the prefrontal cortex, but they also recruit different networks for their particular performance.[10] Those who predicted a single self-contained module—the "will"—may be disappointed to know that the prediction is withering. A network of areas, rather widely distributed, partially overlapping, regulate self-control.

The data converge from a vast number of studies on animals, healthy humans, psychiatric subjects, and neurological patients with brain damage. Obsessive-compulsive subjects, for example, are found to do poorly in the stop-signal tasks, and brain imaging reveals much less activity during the experiment than in normal control subjects in those areas of cortex known to be important for response inhibition. This lower level of activity is quite possibly linked to their distinctive performance. Subjects with trichotillomania (compulsively pulling out one's hair) were even more impaired in response inhibition than obsessive-compulsive subjects. On the other hand, obsessive-compulsive subjects showed reduced cognitive flexibility, as measured by the changes in strategy during a game where the rules changed periodically. That is, obsessive-compulsive subjects were much slower to shift to the new strategy to match the new rules than were trichotillomania subjects. For another example, methamphetamine addicts show reduced capacity for response inhibition in experimental settings and a reduced intensity of gray matter in the inferior prefrontal cortex.[11]

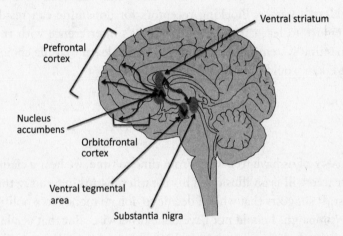

7.1 *Highly simplified illustration of the main components of the reward system, as seen from the medial aspect. Notice particularly the projections between subcortical structures and the prefrontal cortex. Not shown are the amygdala, which plays a crucial role in emotional responses, including learning what to fear, and the hippocampus, which is essential for learning about specific events. Working memory, which also figures in reward system learning, involves the cortex, especially the prefrontal region on the lateral aspect.* Drawn from a Motifolio template.

The subcortical structures include the various components of the reward system (Figure 7.1). Pleasure and pain are associated with dopamine release, which in turn is important for learning which action to repeat and which to avoid. The relevant subcortical structures also include the regions in the basal ganglia that are important for organizing behavior and skill learning, again relying on the reward system and dopamine release.[12] The interconnections between these areas are intricate and difficult to study, so there is much that is not known.

We do know, however, that self-control also depends on the various neurochemicals that modulate the interactions between neurons—neurochemicals such as serotonin, noradrenaline, and dopamine. For example, very high levels of serotonin can increase the willingness to take risks and can even lead to reckless behavior. Blocking receptors for dopamine can result in failure to learn from errors; cocaine's interference with the dopamine system can also lead to failure to learn that the choice not taken would have been better.[13]

FREE WILL

DO ANY of us have free will? From time to time, we hear a claim that free will is an illusion. This is rather disturbing, to say the least. It suggests that when I decide to donate money to a political campaign, I could not have done otherwise. But that is a bit of a stretch. I am not a puppet; I could have done otherwise. So what do these claims mean? Do they mean that you are never in control? That is far-fetched. If a grizzly can exercise self-control, why not you?

First we have to get straight what different people mean by *free will* or the conversation collapses into confusion—as it might if you and I were talking about spare tires and you meant the extra one for the car in case of a flat and I meant the fat around one's midsection, called a "beer belly." Not the same thing at all. Apart from the issue of what we generally *mean* by free will in ordinary contexts, there is the further matter of what in fact *occurs* when a person gathers evidence, deliberates, seeks advice from others, and then comes to a decision. New discoveries about what actually happens may invite us to modify our meanings somewhat, or perhaps a lot, or, depending on the discoveries, not much at all. To take an unrelated example, people in the mid-nineteenth century believed that atoms were the indivisible units of matter. When it was discovered that in fact atoms have structure—protons, neutrons, electrons, and so forth—the meaning of the word *atom* slowly changed to reflect those discoveries. I suppose people might have said, "Well, so atoms are an illusion" and given up the old word in favor of a new one, maybe *dinkytoms*. As it happens, they did not.

There are essentially two completely different things that can be meant by *free will*. First, you can mean that if you have free will, then your decisions are *not caused* by anything at all—not by your goals, emotions, motives, knowledge, or whatever. Somehow, according to this idea, your will (whatever that is) *creates* a decision by reason (whatever that is). This is known as the *contracausal* account of free will. The name *contracausal* reflects a philosophical theory that *really* free choices are not caused by anything, or at least by nothing physical such as activity in the brain. Decisions, according to this idea, are created free of causal antecedents. The German philosopher Immanuel Kant (1724–1804) held a view roughly like this, and some contemporary followers of Kant do also.

So far as I can tell, this is an idea espoused mainly by academic philosophers, not by dentists and carpenters and farmers.[14] When I take our dogs to the beach and ask other folks what they mean by *free will*, no one has this contracausal sense in mind. No one. They doubt it even makes any sense at all, even when I present the idea sympathetically. They do not think conviction under the criminal law requires free will in the contracausal sense. Without the slightest hesitation, lawyers tell me this as well. I am not claiming to have conducted a scientifically valid survey, but my experience with ordinary people is that free will has a different meaning from the philosophical sense of "contra-causal choice."

Still, philosophers who are inclined to favor the idea of a non-physical soul are apt to imagine that the soul acts for reasons, and when it does, then the soul is not caused to decide anything. Rationality, in this view, just works in a cause-free zone, and reasons are not causes.

One problem with this theory is that it elevates having a reason and acting on it to a semimagical level. Such an elevation is unhelpful. A reason can be a perception, an emotion, a memory, a solution to a problem, an evaluation of the future consequences of an action, a judgment about the weight of the evidence, and so forth. *Any* of these can be reasons, and all involve functions carried out by the physical brain, almost certainly.

The related problem is the assumption of a nonphysical soul, a dubious assumption, as we have seen, that has found itself singularly evidence-weak. A third problem is this: When I make a bad decision, such as impulsively running from Butch while trying to deliver eggs, is that decision caused by fear in my soul, or are souls out of the story altogether when decisions are impulsive, such as those caused by overwhelming fear? Are only good decisions created by the soul, while bad ones are caused by the brain? This is not a subtle problem. Unfortunately, any soul-based strategy for solving it seems ad hoc—made up for the occasion rather than part of a systematic and powerful explanatory theory.

Here is the second, and I believe ordinary, meaning of "free will."[15] If you are *intending* your action, *knowing* what you are doing, and are of sound mind, and if the decision is not coerced (no gun is pointed at your head), then you are exhibiting free will. This is about as good as it gets. Moreover, that is generally good enough for practical purposes. We are all familiar with the prototypes of voluntary, uncoerced, intended actions, and we regularly use the categories *intentional* and *voluntary* to draw meaningful distinctions, both in everyday life and in the court-room. I call this the regular (as opposed to contracausal) sense of free will. It is also the sense used in legal contexts. This should surprise no one.

This second meaning is not backed by a precise definition in the way that the word *oligarchy* is: "government by the few." Although many scientific concepts, such as *electron* or *DNA*, are now precisely defined, it was not always so. The science had to mature to the point where it was able to provide such precision. Moreover, other scientific concepts, such as *personality*, *addict*, and *mood disorder* are not very precisely defined, but we com-municate very well nonetheless. So the lack of precision for the term *free will* does not imply that we are muddled; it just means that the concept is like most everyday concepts that we use effi-ciently without fussing too much. Moreover, like *gene* or *protein*, it may become a little more precise as a result of developments in science.[16]

Mort Doran is a highly respected physician and surgeon in Cranbrook, British Columbia, who has had Tourette's disorder all his life.[17] As he explains it, from time to time he has an over-whelming need to repeatedly go through a whole series of mean-ingless actions (touch the nose, touch the elbow, and so forth) and to vocalize words such as "stupid, stupid" for no reason at all. These meaningless actions are referred to as *tics*, and the behavior is known as *ticing*. The sensation of needing to make certain favored gestures builds up and builds up until Dr. Doran goes out behind the house and gives in to the powerful impulses. His wife and sons are completely used to him and do not pay any attention to the tics. Amazingly, when he needs to perform an appendectomy, he does so flawlessly, with no Touretting gestures or expletives at all. He is calm and entirely controlled. After the surgery is done, he walks out of the operating room, and the ticing resumes with renewed intensity.[18]

In terms of *free will* versus *no free will*, how should we think of those periods behind the house when Dr. Doran lets fly? First, we should resist the assumption that either his actions are free or they are not free, either voluntary or involuntary, with no space in between. The fact is, his control functions are just somewhat different. It is not exactly that he is entirely out of control; nor is he fully in control. He is somewhere in between. If a house fire were to break out, he would not just stand behind the house ticing. We all make various little purposeless movements from time to time—a scratch here or there, a sigh heaved, a nail bitten, a leg crossed. One way to think of the compulsion to tic is that it is a bit like those familiar fidgets, only the desire is more intense and the actions more ritualized.

Sometimes during a boring lecture, my desire to leave the lecture hall can become very intense. I have to exert every effort to stay in my seat, pretending to listen intently. It exhausts me. Dr. Doran reports that suppressing his desire to tic is exhausting. Maybe we are not so very different. Perhaps he has just a little more dopamine in the PFC.

The neurobiology of Tourette's disorder is beginning to be understood, and not surprisingly, the areas involved are those in the prefrontal cortex related to self-control capacities and connected to the basal ganglia. Dopamine and noradrenaline neurotransmitter differences are also implicated.[19]

The important point to be drawn from the example of Tourette's disorder is that nonprototypical cases, cases not readily classified as voluntary, abound. Some cases are not momentous, so we ignore them and get on with other things. Some are very serious and are argued in courts of law: the severely depressed woman who drowned her babies, the 10-year-old who shoots his abusive father, the manic-depressive who buys two Jaguars in a single afternoon, the patient in the early stages of Alzheimer's disease who makes a decision regarding treatment.

Although all language users generally know implicitly that

categories such as *freely chosen* have fuzzy boundaries and nonprototypical cases, explicitly emphasizing the point is useful. For one thing, we are released from the desire to insist on precision where none exists. Forcing precision where none exists often means you end up wrangling about words instead of making progress in solving a problem. Additionally, the observation accounts for reservations about how to handle cases in the boundary areas. Does someone with Tourette's disorder have free will or not? Well, sometimes they are in control, sometimes not exactly. Their ticing can be modulated by certain drugs, and they make entirely rational and difficult decisions in much the same way as anyone else.

Moreover, the fuzzy boundaries of concepts such as *self-control* and *voluntary* and *free will* explain why intelligent, conscientious people might disagree about how to judge problematic cases. I am inclined to think that the wealthy actress who was caught shoplifting pricey underwear should be held legally responsible; you think she was in the grip of an irresistible temptation and needs treatment, not punishment. This is a boundary case, not a prototypical case, in contrast to Bernard Madoff and his Ponzi scheme that ran for more than a decade. Note, too, that in the case of the shoplifting film star, there may be no right answer, and so we try to resolve our differences as best we can.

This perspective on prototypes and fuzzy boundaries also implies that while we can hew out rough definitions, getting precise definitions—ones that sort out in a systematic way all the messy cases in the boundary area—may continue to elude us.

In the context of the law, decisions do have to be made. By and large, they are made relying on the sophisticated framework of the law, with goodwill and understanding, in recognition that many cases are tough cases, inasmuch as they do not straightforwardly fall into the available categories. Disagreement may persist about a judgment, however.

Now let's get back to the startling claim that free will is an illusion.[20] What exactly is that claim supposed to mean? It could mean any number of things, depending on the author. If you are a Kantian, for example, you are convinced that a free will must be free of all antecedent causes; genuine choices must be created by the will without causal inputs. Then you come to realize that decision making in the real, biological world always involves causality. So you dump the very idea of free will. You then announce, "Free will is an illusion." In short, you mean that *contracausal* free will is an illusion.

Well, sure. But if contracausal choice is the intended meaning, the claim that free will in that sense is an illusion is only marginally interesting. Because nothing in the law, in child-rearing, or in everyday life depends in any significant way on the idea that free choice requires freedom from all causes. As the philosopher Eddy Nahmias has shown, normally people do not think that free will is to be equated with contracausal free will.[21] Perhaps the claim might be interesting to a small subset of Kantian philosophers. As for the rest of us, it is a bit like announcing with great fanfare that alien abductions are not real. Oh really, no flying saucers arriving on the lawn in the middle of the night? Gosh. Thanks for the heads-up. But what if *free will is illusory* means something else? What if it means, for example, that because there is a neural substrate for our deliberations and choices, we cannot have free will? Now I am totally at a loss. Why would anyone say such a thing? So what do they think *is* required to make genuine choices? A nonphysical soul? Says who?[22]

If *free will is illusory* means that there is no difference between a brain with self-control and one lacking self-control or one with diminished self-control, that claim is flatly at odds with the facts. As discussed above, those differences are entirely real, and quite a lot is now known about the neurobiology of those differences.

With some disappointment, I am bound to say that I suspect

that the claim that free will is an illusion is often made in haste, in ignorance, and with an eye for the headline and the bottom line.

What is *not* illusory is self-control, even though it can vary as a function of age, temperament, habits, sleep, disease, food, and many other factors that affect how nervous systems function. Nonetheless, evolution, by culling out the inveterately impulsive, saw to it that, by and large, normal brains have normal self-control.[23]

CRIME AND PUNISHMENT

WHEN DO we really care about whether someone acted in a voluntary manner? When does it matter? It matters most when someone has caused an injury or harmed another; when someone threatens the stability and well-being of the community. Then the serious questions of intent and control are not merely a matter of academic volleyball. How those questions are answered in a specific case matters enormously—for the accused, for the victim, for the level of trust the community puts in the system, for whether rough justice usurps carefully crafted community practices. This is deadly serious.

One night, our neighbor to the north, Mr. Cameron, got up in his pajamas and set fire to his pear trees with a blowtorch. As this is a desert area, the possibility of the fire spreading caused alarm. What was the motive? There was none that anyone could fathom. It turned out that he had become very demented, a fact carefully kept under wraps by his distressed wife. He essentially had no idea what he was doing. It was not that he was trying to harm or cheat anyone. He was just utterly confused.

A different fire required a different analysis. On a winter night, the Ketterings' barn, badly in need of repair, went up in flames. Kettering had long needed a new barn. When the evidence emerged that he had set the fire himself after removing everything valuable, including the cow, he was charged with arson. He had no excuse from responsibility. He had hoped to cash in on the insurance and get himself a fine new barn. There was motive and there was criminal intent. Yes, he needed a new barn, but he planned and carried out an action he knew to be criminal.

Determining whether someone is responsible for a socially harmful act has long been the focus of thoughtful reflection. In modern times, this reflection has come to be codified in principles of law. Criminal law is profoundly pragmatic, and matters of social safety and protection are paramount.[24] Because intent and self-control are assumed to be the default condition in a defendant—they hold unless the defendant can show otherwise—it is useful to balance the more abstract discussions of free will by drawing on the wisdom of the law concerning how it accommodates departures from the default condition.

Criminal intent is always a relevant matter in determining responsibility for a violation of the law. Despite the many ways in which the question of intent can get complicated, in cases of arson, embezzlement, and fraud, criminal intent is normally quite easy to establish. Long appreciated are the complications that arise when the defendant is not of sound mind. The criminal code has very specific criteria for whether the accused is competent to stand trial: Can the accused participate in his defense, and does he have a rational understanding of the charges against him? For example, a patient with advanced Alzheimer's disease is unlikely to be evaluated as competent to stand trial. If the defendant is competent to stand trial, he may nevertheless be counseled to enter a plea of insanity. The main legal requirement for a plea of insanity concerns whether the accused knew that what he was doing was wrong at the time he did it, often referred to as the M'Naghten Rule.[25] In some jurisdictions, there is also a provision for a plea of insanity concerning volition. This provides for the possibility that even though the defendant knew what he was doing was wrong, he could not prevent himself from doing it.

In the United States, the volitional prong became a focus of dissent in the aftermath of the Hinckley trial, in which John Hinckley was tried for attempting to assassinate President Ronald Reagan. The jury found Hinckley not guilty by reason of insanity, on the volitional prong. Allegedly, his obsession with film star Jodie Foster was so overpowering that he planned to kill the president to impress her. His lawyer argued that he could not stop himself, even though he knew the action was wrong. The public response to the verdict was not generally favorable, and thereafter the federal government and a majority of states abandoned the volitional prong as an option for the defense.[26]

When does a question of self-control in adult defenders arise? A partial defense for murder allows the defendant to argue for loss of self-control under great provocation. This is taken to mean that there was a qualifying trigger, such as overwhelming fear of serious violence, and that a normal person of the same sex and similar age and with normal restraint and tolerance would have reacted in the same or similar fashion. Such a partial defense has been used in cases where a battered wife kills her husband, for example. The previous law on loss of self-control required that the loss be sudden, whereas the updated law that took effect in 2009 drops the requirement that the loss be sudden. While welcome in some respects, this change raises challenges concerning how to distinguish nonsudden loss of self-control from garden-variety revenge. As a more general point, it is reassuring to see that the law aims to be both sophisticated and practical in its sensitivity to circumstances of diminished self-control.

As science explores the causes of violence, the suggestion may be floated that if a defendant can be shown to be predisposed by nature to lose self-control, then he should be exonerated. Providing compelling evidence for such a predisposition has not been straightforward. Recently, however, studies identifying a specific genetic variant have come to the attention of defense lawyers. Several large epidemiological studies indicate that males who have a certain genetic variant for the enzyme monoamineoxidase-A (MAOA) are highly likely to show self-destructive and resilient aggression *if* they were also subject to abuse as children.[27] (Hereafter, I shall use *MAOA x abuse* to refer to this pairing of the variant gene with the environmental factor.) The gene is located on the X chromosome, so the condition would not be seen in females unless, as must happen very rarely, the genetic variant is found on both her X chromosomes. The incidence of *MAOA x abuse* in human females is unknown.

Some features in the self-control networks of males who both carry the variant and are victims of abuse are presumed to be different from those of males with the normal *MAOA* gene. Exactly what in the brain these differences amount to has not yet been tracked down, however. At the behavioral level, the net effect is that *MAOA x abuse* males are highly likely to be volatile, self-destructive, and aggressive given minor provocation, imagined provocation, or no provocation at all.

The main philosophical point urged by those supporting exoneration for defendants in the *MAOA x abuse* category is that neither their genes nor their early abuse were chosen by them. That they are predisposed to violence is not their fault. Of course, none of us chose our genes, so the significance of that part of the argument is dubious. The matter of choosing genes aside, exoneration is a rather extreme recommendation for these cases, one unlikely to be acceptable to victims and their families, let alone the wider community. Precisely because these individuals are unusually prone to violence, even without provocation, releasing them back into society is not obviously a wise policy. How does the law regard *MAOA x abuse* cases?

Two court cases are relevant here. In the first, an Italian case, the defendant, Abdelmalek Bayout, was assaulted by a group of youths. He then bought a knife, followed the victim down the street, and killed him. As it happened, the victim was not among those who had assaulted Bayout. In the sentencing phase, MAOA evidence was produced (the case for abuse was unclear, and psychiatric status was an issue), and his sentence was reduced by one year.[28]

The second case was tried in Tennessee. During a domestic argument, Bradley Waldroup shot his estranged wife's friend eight times and attempted to kill his estranged wife, shooting her in the back and then chopping her with a machete. *MAOA x abuse* data were entered into evidence in the liability (preconviction) phase, and the charge of murder was reduced to voluntary manslaughter.[29]

One important question regarding this sort of evidence is whether the law should specifically accommodate this kind of genetic data, and if so, how.[30] Should it be allowed at all? If so, in what phase of the trial—only in the sentencing phase or in the liability (preconviction) phase? In his careful analysis of the two *MAOA x abuse* cases, Matthew Baum argues that the most appropriate place, if any, for a modification to the law would be to allow the evidence in the postconviction phase, as relevant to mitigation of sentence, as happened in the Bayout case.[31] Baum's general view about changing the law to accommodate these cases, which I share, is one of great caution. For one thing, there are still outstanding scientific questions. For example, one concern about the studies on *MAOA x abuse* subjects is that the populations were unusually homogenous (one in New Zealand, one in Sweden), and it is not yet known whether the link between the genetic variant for MAOA and a predisposition to violence holds more generally.[32]

The report concerning the predisposition to violence in *MAOA x abuse* individuals is apt to raise questions about treatment. For these individuals, so the argument goes, justice would be better served by treating them rather than by punishing them, especially because they have not chosen to be predisposed to violence. Although a useful discussion topic, the treatment option is confined to the world of the hypothetical. At this time, there is no effective treatment for these cases, and it is difficult to predict when such treatment might be available.

More generally, it is important to appreciate that if treatment is to be a viable alternative to custody for a criminal offender, if it is to be an *alternative* that a fair-minded public can accept, the bar for efficacy of the treatment must be set very high: the treatment must be shown to be sufficiently effective as to make reoffending very highly improbable. Not just that it works sometimes in some cases, but that its efficacy has been sufficiently well established so that we can be reassured that the treatment truly is effective in preventing reoffending. This is a strong, but not impossibly strong, requirement. Because offenders often do reoffend, the requirement must be strong.

In a medical context, intervening with a treatment to alleviate symptoms need not meet such a high standard. Prescribing treatment on the basis of clinical trials showing that the treatment is successful more often than not can be good enough. It is not unreasonable to try an intervention and see if it works for this patient. But the clinic is very different from the legal context where a violent crime has been committed and where there is nontrivial risk that a released offender will reoffend.

By and large, there are no ethically permissible psychiatric or pharmacological treatments that can be relied on to render a repeat offense highly improbable. Prefrontal lobotomy, for example, which cuts the connections between the prefrontal cortex and more posterior regions of the cortex, might be an effective treatment, but it is not ethically permissible. Unfortunately, effective, ethical treatments for nervous system disorders are in general very hard to come by. Offenders differ in their backgrounds and personalities, and the neurobiological etiology of violence is poorly understood. This does not mean that offenders should not be treated with what is available, since some offenders may well benefit to some degree, bearing upon their parole, for example. What it does mean, however, is that such treatments are not a substitute for custody. In the future, treatments may be developed that have such a status, but none do now.

Treating sexually violent offenders to curtail reoffending has been an appealing approach in many countries, including Germany, France, Sweden, and Holland. Without treatment, the recidivism rate is about 27 percent; with treatment, it drops to about 19 percent. The treatments range from the pharmacological and surgical (for example, voluntary castration) to cognitive-behavioral therapy. In their careful meta-analysis of available studies, many of which were methodologically nonconforming, psychologists Martin Schmucker and Friedrich Lösel found that there was about a 30 percent reduction in recidivism for pharmacological, surgical, and cognitive-behavioral treatments.[33]

Castration was found to reduce recidivism in violent sexual offenders to about 5 percent, though Schmucker and Lösel point out that those who elected to undergo castration were exceptionally motivated to avoid reoffending. This already set them at a much lower recidivism risk than the population of sex offenders in general.[34] Offenders receiving hormonal treatments to reduce testosterone levels (antiandrogen treatments) also showed significant reduction in recidivism. A practical problem with the antiandrogens is that when individuals are released from custody, they may stop treatment because they dislike the side effects, including weight gain. Unfortunately, being on and then going off antiandrogens tends to result in an even higher recidivism rate than baseline. Even for castration, moreover, later applications of exogenous testosterone can reverse the effects. For both these treatments, Schmucker and Lösel emphasize that the results should be considered very cautiously, since the lack of uniformity in methods across studies makes meaningful comparisons problematic. From a legal perspective, these studies are most relevant to the question of treatment during and after custody, rather than as a substitute for custody.[35]

Different considerations arise concerning offenders who have a tumor compressing the orbital region of the prefrontal cortex, a region known to be important for self-control.[36] A conscientious defense attorney may want to argue that the offense was owed to the tumor and that removal of the tumor will reduce the risk of a repeat to zero. An image of such a tumor can be very dramatic and may have the potential to influence a jury whose members lack the scientific expertise to evaluate the significance of the image they are shown.[37] As things stand, the propriety of admitting such images into evidence remains highly contentious. This is largely because the science cannot establish that the condition was causally determinate in the person's decision making at the time of the crime.

From a medical standpoint, it may seem probable that the tumor had some effect in the defendant's overall cognitive functioning. From a legal standpoint, however, the question to be addressed is more precise: Did the tumor cause the defendant to strangle his wife?[38] An obstacle to demonstrating this claim is the fact that of the people who do have such tumors, very few commit a violent crime. This fact detracts from the cogency of the specific claim that this particular violent action (strangling the wife) can be explained as owed to the presence of this particular tumor. It implies that other factors, presumably unknown, must have been significant. Because so much remains unknown about the details of the circuitry regulating self-control and the emotions, and because individual differences in the microcircuitry make tracking causal relationships in an individual case extremely difficult, judges and prosecuting attorneys prefer to err on the side of caution.

My point in dwelling briefly on the matter of treatment is to draw attention to the practical realities of treatment for violent offenders. The regrettable condition of prisons often motivates well-meaning people to question why treatment is not used instead. Unfortunately, however desirable the ideal of successful treatment may be, the science cannot yet deliver a treatment of sufficient efficacy to put the ideal within reach. Public protection, of necessity, will always be a major factor in the workings of a criminal justice system, and until treatment can reduce recidivism to an extremely low and publicly tolerable rate, custody will continue to play a major role. I hasten to add that these issues are very complicated, and there is much more to be said on this topic.

Scientists have an enormous obligation to appreciate the consequences of supplying their expertise in a legal context. Slapdash a priori arguments that no one should be punished for anything are not only disingenuous, but tend to discourage productive discussion. Empty the prisons? Vigilante justice would immediately take hold, bringing us far more misery than even an imperfect criminal justice system. It is folly to suppose that punishment would cease just because the prisons were emptied. It is folly to suppose that the criminal law, with its threat of punishment, fails to deter individuals from doing terrible things. Suppose there were no penalties for cheating on your taxes. Would most people nevertheless faithfully pay their owed amount, or would they shave a bit here and there or perhaps let payment quietly slip their mind altogether? Just as attorneys and judges need to learn about those aspects of neuroscience that will help them understand and argue a case, so scientists and philosophers need to understand a little of the history of the criminal justice system, the defining cases, and the arguments that sustain judicial practices.

SOMETIMES THE impulses you suppress are impulses to believe. Balance and judgment are emotional-cognitive elements in your mental life, and you rely on balance and judgment to say, "No, I am not convinced that apricot pits will cure ovarian cancer" or "No, I am not convinced that a self-proclaimed psychic can find the lost child" or "No, I am not convinced that my investments will make 12 percent per year for the next 20 years." You hope that misfortune will not happen to you, but you take out insurance against that possibility. You hope that you will never take a nasty fall while skiing, but you wear well-fitting boots and a helmet in case of accident. No parent wants to believe that their child is a cutup in French class, but it may be best to face the facts anyhow.

As with all things biological, there is great variability across individuals of a species. Some are chronically gullible, others hamstring themselves with suspicion; some are reckless, others are excessively cautious. Self-control plays a role in what you believe every bit as much as it does in what you decide to do. The habit of questioning and taking a second look is often a survival advantage. Under certain conditions, however—perhaps in the heat of battle—acting on impulse may be more advantageous than taking a second look.

One issue that I left on the shelf in this chapter concerns the role of unconscious activities in our decisions and behavior, including highly controlled behavior. The unconscious brain is the focus of the next chapter.

Chapter 8

Hidden Cognition

UNCONSCIOUSLY SMART

Suppose you are meeting your boss's husband at a cocktail party. Here is something you are going to do—unconsciously. Once you are introduced and begin chatting, you will tend to mimic the smiles, gestures, and speech intonations of the man. Ditto for him with regard to you. He takes an appetizer, then so do you; he crosses his arms, and a few seconds later, so do you. You use an exclamation such as "remarkable!" and after a few seconds, he echoes "yes, remarkable." This subtle and sophisticated unconscious *mimicry* is especially common when two individuals meet for the first time, but is certainly not restricted to those occasions. We *all* do it all the time, and not just at first meetings. Careful observations by experimental psychologists show that unless the context is ominous, two persons regularly and subtly mimic each other's social behavior. Yes, you do it, too.

Usually, this sort of mimicry is not done with conscious intent. That is, you are unaware of intending to do it or even that you are doing it. Moreover, it is sensitive to context. The greater the importance of making a favorable impression or the more socially stressed you are, the more likely and frequently you will mimic. Not only are you generally unaware of your behavior *as* mimicry, the corresponding low-key mimicry on the part of the other person can escape your conscious notice, too. It nonetheless makes a difference to how you size up each other.

What is the evidence that you, I, and most everyone else does this pretty regularly? Social psychologists have run scores of studies with this sort of protocol: An undergraduate is seated at a table in a room and asked to work with another person (actually the experimenter's helper) on some puzzle or other innocuous task. Hidden observers count the instances of mimicry by the student. Videos make the mimicry plain. If the student is socially stressed before the trial, mimicry increases.

You can vary the protocol to determine the effect of mimicry or lack of it on the student. In this variation, the helper either mimics the student's gestures, body posture, and so forth, or carefully suppresses mimicry. After leaving the room, the student is asked to evaluate the helper. If the helper mimicked, he is probably liked; if he suppressed mimicry, he probably is not liked. Here is another way to see the effect. If the helper apparently inadvertently drops a cup of pencils, the mimicked student is more likely to help pick them up than is the nonmimicked student. Asked later whether they noticed that their gestures were or were not copied, the students have no clue.[1]

Of course, the mimicry *can* be performed with deliberate intent as well, and sometimes salespeople are instructed in the fine art of just-right mimicry to smooth the way to sales. They are drawing on the fact that we feel more comfortable when we are mimicked. Psychiatrists have also told me that they often deliberately mimic just enough to encourage their patients to relax. (They might drop the "g" and say, "How are you doin'" if the patient isn't a college graduate.)

Why do we do it? Social psychologists believe that this sort of mimicry in social situations tends to enhance the trust each has with the other. If I do not mimic you at all in our conversation, you are apt to become rather uncomfortable and to find me a little less *sympatico* than if I do mimic you. But *why* does it make us more comfortable? There is no established theory on this. My own speculation runs this way: If, albeit in this very low-key way, you are behaving like me, then in some significant social respects you are similar to me; if I know that, then you are less unpredictable to me. You are like those in my tribe, my clan. And brains *love* predictability.[2]

You may also find that you cannot easily prevent yourself from engaging in social mimicry unless you exert considerable effort. Moreover, if you do succeed in suppressing mimicry, you are apt to be socially quite uneasy about the success. You may have made the other person feel disliked; you may have made yourself disagreeable. If you are in the military or the CIA, you may be trained to control this sort of social behavior in the event you must negotiate a very delicate situation.

Time for a brief *whaddyamean* sidebar. The words *unconscious*, *nonconscious*, and *subconscious* might have subtle differences of meaning as used in various subfields of science. Nevertheless, what is common among them is that the processes referred to are not ones of which we are aware; we cannot report on them, though they can influence our behavior. This is not a definition in a tight sense, since tight definitions go hand in hand with developed science. In the *conscious/unconscious* case, brain science has not developed far enough to say exactly what is going on when we are aware of something and when we are not.

What we do while awaiting scientific results that will float a tight definition is refer to prototypical cases, those where agreement is standard. For example, it is generally agreed that you are unaware of the processes whereby you see a face as a face or whereby you know you are about to vomit or sneeze; you simply are aware that you are about to vomit or sneeze. You are unaware of sensory stimuli when you are in deep sleep or in coma or in a vegetative state. You are aware of seeing faces and smelling odors when you are awake. You are unaware of the processes whereby you know that someone misspoke himself. So although you might suspect I should carefully distinguish between unconscious, subconscious and nonconscious, for the purposes at hand it would be like polishing a rat trap—not necessary to get the job done.

YOUR UNCONSCIOUS DOES THE TALKING

THINK FOR a moment about your last conversation. Did you consciously choose your words? Did you consciously organize the structure and word order of the sentences you spoke? Did you consciously avoid using words that you know would offend your listener? Almost certainly not. Normal speech is under the guidance of nonconscious mechanisms. You became conscious of precisely what you unconsciously intended to say only when you said it. You modify your speech depending on whether you are talking to a child, a colleague, a student, or the dean. Not consciously, most probably.

Paradoxically, speech is usually considered the paradigm case of conscious behavior—behavior for which we hold people responsible. Certainly, it does require consciousness; you cannot have a conversation while in deep sleep or in coma. Nevertheless, the activities that organize your speech output are not conscious activities. Speaking is a highly skilled business, relying on unconscious knowledge of precisely what to say and how.

Freud, though perhaps best remembered for his ideas about psychoanalysis, had theories about consciousness that are arguably much more important than his rather tendentious ideas about curing neuroses, theories that he developed around 1895. The young Freud was a neurologist who specialized in the loss of language capacities (aphasia) following stroke or other forms of brain damage. Freud realized that when we talk to each other or give a lecture or an interview, we are conscious of what we are saying, but the selection of words and the organization of words into sentences and arguments are all managed by unconscious processes. We are consciously aware of the general gist of what we want to say, but the details come out of the unconscious brain. Most of the time, this "general gist" may itself not be articulated in words. It is represented only in a vaguely imagistic way. Sometimes we have at best only a semi-awareness of what we intend to say.

You will have noticed that if you stop to consciously prepare precisely what you will say next, you become tongue-tied. Then you do not talk in a normal fashion at all. I am sometimes amazed that as long as I know the general outline of what I plan to say in a lecture, it just all flows out, more or less as it should, but the word-by-word planning is a job my unconscious brain handles for me. When I first gave public talks, I tended to overprepare, trying to have exact sentences committed to memory. This was a disaster, as it sounded stilted and lacking in spontaneity, which indeed it was. Then I began to trust myself to express the knowledge I knew I had—the gist of what needed to be said—and the words just came, mostly in an acceptable form.

Here is what I find amazing when I give a lecture, given the role of the unconscious brain: I virtually never say anything completely irrelevant and out of context. Though I have a rather bad habit of using cusswords when among friends—"bloody this" and "bloody that"—when I give a lecture, I never say "bloody" anything. I do not even have to think about not cussing. Rock on, unconscious brain.

Lots of people will have noticed this aspect of speaking, but Freud went on to ask an interesting question: Are those unconscious processes *mental*? He thought, *yes*. Freud's 1895 claim that an intention could be both unconscious and mental caused a furor. That is because in the 1890s, many scientists, as well as most people generally, were dualists. In short, they believed that mental states such as hearing the dog bark and thinking about Mount Fuji were states of the nonphysical soul, not the physical brain; after all, they are *mental* states.

Freud's neurology colleagues agreed that the processes supporting speech were, to be sure, brain processes, but they insisted that such processes therefore could *not* be mental. Instead, these processes were understood to be biological and hence more like reflexes than like intelligent problem solving.

Freud stood his ground. One reason he thought that the processes supporting speech were both mental *and* physical is this: When you and I are talking and you utter a particular sentence, presumably you intended to say what you said, yet you did not formulate that sentence first in your conscious mind and then deliver it in speech. So in a certain sense, the intention was unconscious. But your comment is nothing like a reflex such as the startle response evinced when someone sneaks up behind you and hollers.

According to the conventional wisdom of Freud's time, the vocabulary appropriate to explaining unconscious business encompassed words like *neuron*, *reflex*, and *cause*, whereas the vocabulary appropriate to explain mental business encompassed words like *intention* and *reason*. And never shall the twain vocabularies meet or mesh. Brains cannot reason, minds can; brains have neurons, minds do not. According to the conventional view, the very idea of unconscious perceptions or thoughts or intentions was paradoxical. It made no sense, because it used language in an unconventional way. Freud saw things more deeply. The processes underlying speech are not, *not by a long shot*, reflexive in the way that a startle response is reflexive. They are smart and appropriate and intentional. They are unconscious. And they must be states of the brain, because there is nothing else for them to be states of.

Like Helmholtz, Freud came to the realization that dualism of the *mental versus the biological* was misguided. What weighed against dualism were all the problems that have tormented dualism since Descartes, problems we discussed in Chapter 2, such as: What is the nature of the relation between physical stuff and soul stuff—how can they causally affect one another? If souls have neither mass nor spatial extent nor force fields, how can brains, which have all these, interact causally with such a substance? Moreover, where would the soul come from, assuming Darwinian evolution? When does the soul enter the body, and how does it do that? How would the existence of a soul affecting and being affected by the brain square with physics and the conservation of energy and momentum? Why, when the brain loses tissue, is memory diminished, if memory is part of the nonphysical soul?

Reconstructing and simplifying, I think that Freud realized that what we refer to as the mental *is* neurobiological, however much this might jostle our intuitions. He understood that unconscious reasoning and intentions and thoughts need to be invoked to explain such things as complex perception (for example, heard speech as having a specific meaning) and complex motor acts (for example, speaking intelligibly and purposefully). Moreover, he suspected that what was needed in the long, *long* run was a single, integrated vocabulary spanning both conscious and unconscious business. Ultimately—and that could be a very long "ultimately"—that vocabulary would reflect advances in the sciences of the brain and behavior.

In the meanwhile, thought Freud, because the brain sciences are scarcely developed at all (remember, this is 1895), we make do with the vocabulary we have. He realized that he had essentially no idea what a vocabulary spanning the brain and behavioral sciences would look like. His conclusion was that we have no choice but to make do with what we know is a flawed and misleading vocabulary, namely, that of intentions, reasons, beliefs, and so on, to describe unconscious states. Thus, recognizing that we have to manage with the science we have, we can see the merit in characterizing the intentions shaping the exact form of what we say as *unconscious intentions*. This, in a nutshell, is the basic argument Freud put to his colleagues.

On this matter, Helmholtz and Freud won the day, and from their insights emerged new puzzles: What are the mechanisms whereby *un*conscious intentions become *conscious*? Here Freud began to speculate, but in a surprisingly modern way. For sensory perception, he thought, there need be no unconscious intentions guiding what becomes conscious. Rather, the results of sensory processing become conscious as a function of the way the brain is organized. If you open your eyes, you will consciously recognize faces, dogs, cars, and so on. If you attend to the sound of the engine, you will hear it; if you attend to the sound of the raven, you will hear *it*; and so forth. No conscious work—usually, anyhow—needs to be done to realize "oh yes, that is the face of Winston Churchill."[3]

For thoughts, however, Freud seems to have surmised that they have to be "pushed through the language networks" to become conscious. The model for his surmise was speech: I only know exactly what I want to say when I have said it. Similarly, I only know what I think when I say it (silently) to myself. Freud may have been misled about the necessity of language networks for conscious thoughts. After all, quite a lot of conscious thought can take the form of images—sensory, motivational, emotional, and movement images.

Not all reasoning has a languagelike form. We know this from the problem-solving behavior of intelligent animals such as ravens and elephants. As Diana Reiss has shown, elephants can recognize themselves in a mirror, as can dolphins.[4] Bern Heinrich has shown that ravens can solve a novel, multistep problem in one trial.[5] These achievements involve figuring something out, and in my book, that is thinking. Arguably, complex conscious images (the nonlinguistic thinking) are more fundamental for thinking than those "pushed through the language networks." Of course, if you dogmatically define thinking as "must be in language," then it follows trivially that you need language to think. That is really boring.

Here is another reason to believe that thinking and problem solving need not always be language dependent. Preverbal children rely on sensory images and movement images to solve problems and make inferences, such as where a toy is hidden. But with the acquisition of language, children's thoughts can become more complex, reflecting the complexity of the local dialect. Eventually, they learn to suppress overt speech, and the child's covert speech, along with sensory and motor images, constitutes a portion of her conscious thoughts. And of course Freud believed, rightly, that all of this is done by neurons—neurons in networks that collectively do complex jobs.

I also suspect that Freud drew a sharper line than is appropriate between language and other forms of behavior. If you are playing a fast game like hockey or basketball, you may not know exactly what you are going to do next until you do it. You merely have a rough intention. When I am making soup, I have a rough intention of what will go in it, but as I proceed to root around in the refrigerator and the cupboards, other things may end up in the soup. I suppose you could say that I never know what soup I am going to make until I have made it.

Yes, language is important to human cognition, but some philosophers, such as Dan Dennett, have given it what, from the perspective of the brain, is an exaggerated importance. It is as though he took Freud's proposal about the link between thought and language and then extended it to all aspects of conscious experience. Dennett, for example, has long argued that only those with language are actually conscious:

> I have argued at length, in *Consciousness Explained* (1992), that the sort of informational unification that is the most impor-

tant prerequisite for our kind of consciousness is not anything we are born with, not part of our innate "hardwiring," but in surprisingly large measure an artifact of our immersion in human culture.[6]

The part of human culture that is pertinent here, he makes crystal clear, is *language*. The underlying conviction is that consciousness is essentially a narrative, and for a narrative, you need language. Of the cognitive organization (the learned language) needed to support consciousness, Dennett claims: "It is an organization that is swiftly achieved in one species, ours, and in no other."[7]

Dennett is not being cavalier here. He means what he says. He means that without language, an animal is not conscious. That includes nonlinguistic humans. In his 1992 book, he argues that acquiring language rewires the brain for consciousness. This is a very strong and testable claim, for which no evidence from neuroscience has so far been produced. In this conviction, Dennett contrasts starkly with neuroscientists such as Antonio Damasio and Jaak Panksepp, who see other mammals as experiencing emotions, hunger, pain, frustration, and hot and cold, for starters.[8]

This issue will arise again in the next chapter, where we look at progress in understanding the brain processes that support and regulate consciousness. To sow a preliminary seed of skepticism regarding Dennett's idea, I wish to point out, following Panksepp, that a neurobiological approach involves studying the activity of the brains of mammals, including those of humans, when they are awake, in deep sleep, dreaming, in coma, having a seizure, under anesthesia, paying attention, and making decisions. If the brains of mammals under these varying conditions show strong similarities, then it is more probable that the basics of consciousness are to be found across those mammalian species where the similarities to humans obtain.

Birds, whose brains are a little different in their organization, can also be tracked in these ways.

As Panksepp sizes up the issue, being conscious enables the acquisition of language, not the other way around. If you are not conscious, in any of the various ways that a person can be nonconscious (for example, in deep sleep), you are not going to learn much of anything, let alone language.

GETTING THE HABIT

HABITS ARE another way in which your conscious and unconscious brain activities are seamlessly integrated. Forming a habit, whereby much in a job such as milking the cow can be shifted from your conscious brain to your unconscious brain, is a very efficient way of navigating the world. Of course, you have to be conscious to be milking a cow. It is just that an experienced milker can pay less attention to the task than a novice. Habit formation is a godsend. You strive for this in golf, bicycle riding, changing diapers, and shelling peas. It allows your conscious mind to work elsewhere.

Automatizing certain behaviors by acquiring habits makes the business of getting on in life smoother and more polished. In the social world, you acquired habits as a child that serve you well in getting along; now you don't have to think about saying *please* and *thank you* or about not breaking wind in public or staring at someone who is disfigured.

Skilled schmoozers know how to "work the room"; they do not have to consciously figure it out each time. Skilled actors effortlessly "get into the zone." Skilled race car drivers say it is essential not to think (consciously) too much.[9] When you have successfully performed a skilled act, say, playing a Bach partita on the piano before an audience, you feel, "Yay *me*! I did it." Yes, *you* did, you and both your unconscious and your conscious operations. Your unconscious habits are indeed part of *you*. You rely on them constantly; you are relieved when they smoothly operate to get things done.

Some unconscious habits we do not love, such as revisiting time and again and *again* a past confrontation, fantasizing about what we should have said, will say next time, and so forth. Even when we acknowledge the futility of the exercise and consciously resolve to stop going over the event, when we are making the salad or driving to work, along comes the confrontation memory with its compulsive shoulda-coulda-woulda plus the fantasy "add-ons." Our unconscious sometimes seems to behave badly.

Some unconsciously controlled habits might be changed with conscious effort. You might recognize that you tend to react with a strong negative emotion to a certain colleague's predictably long-winded speeches during meetings. I once read that you can control your impatience, as the colleague drones on, by imagining him as a tiny bug in front of you, which you slowly and deliberately capture by imagining that you place a drinking glass over him, as you might with a real glass on a real bug. I have tried this tactic and found it sufficiently distracting (and yes, satisfying) to get through the tedious oration while remaining entirely calm.

Acquiring habits and skills that can be performed pretty automatically is something all brains have evolved to do; it saves time, and it saves energy. Saving both is critical to the survival and well-being of all animals. The *me* that you are depends on a close knitting together of both your conscious and your unconscious business.

ME AND MY UNCONSCIOUS BRAIN

I WONDER: Does the *me* that I am include all that unconscious stuff? Or only the conscious stuff? I think it must include much more than just the conscious events. The brain's conscious and unconscious activities are massively interdependent, enmeshed, and integrated. You would not be who you are but for the well-tuned unconscious business and its tight fit with your conscious life. But for that unconscious business, you would not have a conscious life. You could not remember anything in your auto-biographical past, for example. You could not recognize "Cold, Cold Heart" or Queen Elizabeth. You could not distinguish horses from elephants, because that ability relies on learned differences and unconscious recovery of that learning. You could not distinguish *me* from *not-me*.

Your conscious brain needs your unconscious brain, and vice versa. The character and features of your conscious life depend on your unconscious activities. And of course, conscious events can in turn have an effect on unconscious activities. For example, if you consciously try to remember the layout of your grandmother's house, this triggers unconscious operations of memory retrieval. Suddenly, you have vivid visual imagery of the fireplace in the living room and the dining room off to the side.

Sometimes, while listening to books on tape to ease the boredom of driving, my friend Colleen will make the final turn onto her home street and realize with a shock that she has no recollection of events during the last 30 minutes. She must have changed lanes, responded to the brake lights in front, taken the right exit, but how? Here she is in front of her house, remembering clearly the details of *The Spy Who Came In from the Cold*, but with no memory of the journey. Was she unconsciously driving? Probably not. Most likely, her conscious attention did fleetingly shift back and forth between the driving task and the story, but later, she remembered what was most salient—the story. So she probably did not do the whole drive unconsciously, but some of it she did.

When you are relaxed and spontaneous, on a Saturday morning chatting with the family or sitting around a campfire on a summer evening, you may feel that *that* you is your *real* you. You do not have to be on guard or careful. You can be yourself. Surprisingly, perhaps, even your spontaneity relies pretty heavily on your unconscious brain, as you talk, laugh, gesture, and so forth. So the *real* you, the *you* of Saturday morning at home, is deeply integrated with your *unconscious* activities. You are a highly integrated package—not always a highly *harmonious* package, but a package where conscious and unconscious states jointly contribute to your behavior.

Something strikes you as funny, and with that unmistakable upwelling of feeling, you laugh. Different people find different things funny, and in some way we do not understand neurobiologically, what you find funny is a reflection of you—your age, your past experiences, your personality, as well as various contingencies at hand. When John Cleese in a *Fawlty Towers* sketch becomes so annoyed with his stalled car that he rips a branch from a nearby tree and beats the car, the crowd erupts with laughter. Ask yourself (if indeed you did laugh): Did you consciously decide that this is funny? Almost certainly not. You will be laughing before you can begin to say what makes the scene funny. If you consciously decide to laugh, it is forced and not as enjoyable as spontaneous laughter. What provokes your spontaneous mirth is begun by your unconscious brain, which, incidentally, will be sensitive to whether in the given circumstance it would be rude or improper or dangerous to laugh.

BACK TO THE *ME VERSUS NOT-ME* DISTINCTION

WHEN YOU are on a ferry looking out at the pylons of the dock and ever so gently the ferry begins to move, you may have a fleeting few seconds when you think the dock is moving but you are stationary. An illusion that is quickly reversed, this experience raises a tough brain question: What mechanisms allow the brain to know which motion is self-caused and which not? As we saw in Chapter 2, solving the "what is moving" problem requires a lot of neural sophistication. Even creatures simpler than humans, such as a fruit fly that aims to land on a twig swaying in the wind, need to know what movement is caused by what (me or that) and how to adjust its flight accordingly. An owl aiming to catch a dodging rabbit or wolves cooperating to bring down a fighting caribou have an even bigger problem.

The first point for *self versus nonself* is that solving the "what is moving" problem is very, very basic in the evolution of animal nervous systems. The second point is this: if a brain can solve the "what is moving" problem, then it can make a fundamental distinction between *me* and *not-me*. Other elaborations can extend from there.

The platform for the solution to the problem is this: movement-planning signals in one part of the brain are looped back to sensory and more central regions of the brain, telling them, in effect, "I am moving this-a-way." So when your head moves, your visual system knows to cancel visual motion—nothing out there is moving. Absent such a signal, the motion detected is taken to be motion of *not-me*—of something out there. So when the ferry gently glides out of its slip, you do not initiate the movement, and your brain wrongly concludes that the dock is moving, not you.

As discussed in Chapter 2, the movement-planning signal that gets looped back is usually called an *efference copy*—copy of a *my-movement* signal. The advantage of this separation of signals (*me moving* versus *that thing moving*) is of course that it allows the animal to better anticipate events in its world, thereby having a better shot at self-maintenance. While the logic of the system is straightforward, tracking down the neurons carrying efference copy signals has been anything but, and many questions remain open.[10] The main point in this context is that all this separating of the source of movement is handled without conscious intervention. You just know.

An additional element in the sense of agency—that I was the one who did that—concerns whether consequences of the movement were predicted. If the outcome was totally unpredicted, then I am apt to think that the event was not caused by me.[11] One important way in which the outcome can be unpredicted is that the timing of events is off; that is, the consequences occur much too late or much too early relative to what was expected from past experience. If you hit a cue ball and with much delayed timing the red ball drops in the pocket, you doubt that anything you did made the red ball drop.

You cannot tickle yourself. Why not? The answer is that your brain generates an efference copy that loops back to the sensory systems identifying you as the source of the tickling movement. How do we know that? Here is some evidence. If you rig a device so that when you move a handle, a feather tickles your foot, your brain knows it is you. However, if you put a delay in the mechanism—a few seconds delay between moving the handle and the feather action—then you can tickle yourself (assuming you are ticklish at all, but that is a different story). Basically, with the delay in place, you fool your brain into thinking that the movement is not yours. This means that the brain is very sensitive to the timing of the "my-movement" signal.

When you say out loud, "I need to buy milk," then a signal leaves the movement-planning brain and loops back to the sensory brain to indicate the source of the sound. When you merely *think*, "I need to buy milk," this is covert speech (inner speech). Again, a movement-planning signal informs the sensory brain about the source of the covert speech—*me*. Except sometimes the mechanism is bungled for reasons that are still obscure. So sometimes a person may *think*, "I need to buy milk," but because there is no efference copy signal or none with the right timing, he may fail to realize that his thought is actually *his* thought.

He may in time come to believe that the FBI has put a radio in his brain and is transmitting voices into his brain. Auditory hallucinations are a not uncommon feature of schizophrenia, and a leading hypothesis to explain them refers to the failure of the mechanism for efference copy—identifying yourself as the source of your thoughts. Perhaps the timing of the mechanism, which needs to be very precise, is imprecise.

Given what is known about the neurobiology of efference copy and its relation to a sense of *self*, neuroscientist A. D. (Bud) Craig is exploring a hypothesis about the mechanisms underlying consciousness. Craig suggests that just as the brain uses efference copy to distinguish *movement-out-there* versus *my-movement*, so there is a higher or more abstract level of brain function that, via efference-like loops, distinguishes between mental states that represent *changes-in-my-body* (the *bodily-me*) versus mental states that represent *changes-in-my-brain*, such as being aware that I am feeling angry (the *mental me*). My knowledge of the *bodily-me* concerns such things as knowing that I am running or vocalizing. My knowledge of the *mental-me* underwrites my knowing that the pain I feel is *my* pain, that the fear I feel is *my* fear. This allows me to have a grasp of what I know—for example, that I know the words to "Jingle Bells"—or to have a grasp of whether I am a generous person or a tightfisted person.

Normally, this is all so smooth that it may be surprising to learn that it is necessary to have a mechanism to ensure that one part of the brain knows what the other part of the brain is planning. But the catastrophic confusion in schizophrenia makes us clearly aware of what happens if the mechanism breaks down. Incidentally, it has been claimed that many schizophrenics can tickle themselves, another small piece of evidence in favor of the theory that attributes auditory hallucinations to imprecise timing of the efference copy signal.

A malfunction of the efference copy mechanism, perhaps owed to dopamine imbalances in the brain, may also explain a related phenomenon—so-called automatic writing. A subject may begin writing, convinced that his hand is being directed by someone else because, so far as he can tell, he is not intending to write anything. The writing on the page just happens. Seemingly, the writing does not reflect the subject's own thoughts and ideas; it is not purposeful. So it must be the result of a spirit or alien or ghost.

DID *I* DO THAT?

PETER BRUGGER is a neuropsychologist who recounts the remarkable story of Ludwig Staudenmaier (born 1865), a career chemist with a religious background who began to experiment with automatic writing at the suggestion of a friend.[12] This was during the heyday of spiritualism, and many people sincerely believed they might communicate with dead relatives by letting the hand hold a pencil while the spirit took control of the hand. Why spirits chose this odd route to communicate was not deeply probed. Probably some of those who tried automatic writing helped the process along a bit, not unaware that the spirit was doing less than what was expected.

At first Staudenmaier found the exercise somewhat amusing, but later he became convinced that in fact a spirit or alien being was indeed communicating through him, taking control of his brain and writing thoughts that Staudenmaier himself was surprised by. In his autobiographical book, published in 1912, he recounts the early days:

> After only a few days, I already felt a peculiar pull in my fingertips seemingly aimed at moving the pencil obliquely from left to right. This impression became more and more distinct. Holding the pencil as softly as possible, I concentrated my thoughts on this pull and gave way to it. I attempted to assist

and reinforce it. Within the following two weeks, this process became easier and easier to accomplish.[13]

The idea of communicating with the spirit world was strongly appealing, provoking Staudenmaier to spend even more time in this pursuit. Not very long after the automatic writing was well under way, however, it was replaced by automatic hearing, the writing no longer being necessary and discarded because it was slow. He reported that various ghosts and aliens would talk to him, and he gave them specific names. In his lucid periods he was inclined to attribute these experiences to the brain, for as a well-trained chemist, he did appreciate the bizarre nature of his attribution of the voices to an alien.

Nevertheless, as time went on, the delusions were sufficiently powerful that the skepticism lapsed and he slipped into believing that he really was communicating with ghosts and aliens and so forth. Visual and olfactory hallucinations began to accompany the auditory hallucinations, as when he would see the face of a devil, hear him laugh, and smell sulfur. He later acquired the conviction that he could move objects with his mind, and from time to time he reported that cups flew through the air, presumably tossed by himself. Eventually, he became quite delusional, imagining that there was a poltergeist inhabiting his rectum, a nasty poltergeist who forced him, when feeling cold, to kick his right ankle with his left foot. A different poltergeist, known as Roundhead, sometimes inhabited his mouth, forcing him to grimace and say things he consciously wished to inhibit.

Given the classic nature of Staudenmaier's symptoms, it seems most probable he succumbed to schizophrenia. Brugger is not, of course, suggesting that Staudenmaier's early dallying with automatic writing *caused* the delusions, but rather that his dallying was more likely a symptom of the onset of the disease. Conceivably, however, it might have hastened the onset of his more serious incapacitating delusions. In Brugger's neuroscientific hands, the case is especially fascinating, as it provokes him to explore more widely and to consider that claims about the presence of ghosts causing noises and of poltergeists moving objects are often manifestations of a dysfunction of self-nonself mechanisms. The ghosts are none other than the person himself, unbeknownst to him. Thus, Staudenmaier's own voice is misattributed to the devil or to a ghost; and when an apple flies across the room, he denies any role in the apple throwing. "Not me, must be the devil," he says. Misattributions of self-activity to others seem astonishing and are hard to imagine vividly, which reflects only on the deep-seated, smoothly functioning nature of our own unconscious processing. These misattributions of self-activity to other-activity quite likely involve neuromodulators such as dopamine and serotonin.

In such a condition, the person has no conscious access to such physiological features as mistiming of efference copy or imbalances in his dopamine distribution. His convictions are powerful because of what is delivered to consciousness: that is not *my* voice, *I* did not throw that apple. The breakdown of self-nonself boundaries has been described by schizophrenics once their florid phase has passed, and we do know that such a boundary breakdown causes massive disruption in all aspects of behavior.[14]

His interest sparked by the Staudenmaier case, Brugger then wondered about a connection to a phenomenon more fashionable now than automatic writing, namely, facilitated communication. According to the claims of facilitated communication devotees, autistic people are not really handicapped in terms of language or intelligence; they just cannot communicate. Give them the means, and they will communicate. Parents with severely autistic children are understandably desperate to find some way—any way—to communicate with their child. Moreover, they are susceptible to conspiracy theories that depict in chilling terms the suppression by the medical profession of those with special powers. There are individuals who call themselves communicators who claim to have a special gift or power to facilitate communication with the severely disabled. They do not claim to understand their unique powers, only that they have them. They offer to be the special intermediary between the child and a keyboard, thereby allowing the autistic child to communicate.

The child lays his hand on a keyboard, and the communicator lays her hand on top of the child's. Key-press by key-press, messages emerge such as "I love you mommy," "Yes I can talk." Parents are overcome with emotion.

Can we believe in facilitated communication? Why does the facilitator have to have her hand on that of the child? Here is one test of its efficacy. Show the child one picture (boat) and show the facilitator a different picture (sandwich), and see what answer gets typed out on the keyboard. Be sure the facilitator cannot see what the child sees. Result: the answer corresponds to the facilitator's picture (sandwich), not the child's picture. Repeat. Same result.[15] This and related experiments have been performed hundreds of times, on 23 different facilitators, and the response invariably reflects what the facilitator saw, not what the child saw. Never once did the communicator report the picture shown to the child. Moreover, the children often are not even looking anywhere near the keyboard and essentially have no idea what keys are being typed in.

The test results have been unequivocal and devastating. The countervailing data notwithstanding, the faithful remain believers, and millions of dollars of public money have been spent hiring facilitators at schools. Of great concern also is that facilitators are being used for patients in coma as well as autistic persons in order that the patients can tell us what they want—to go to a nursing home, to stay with their parents, and so forth. These expressions are actually those of the facilitator, not the patients. For families, the situation is grim; they want to believe, but the data completely undermine belief in facilitated communication. It is not real.

Are the facilitators just perpetrating a hoax? Is it that simple? In many cases, the facilitator seems to genuinely attribute the choice of keys to the child and to completely lack awareness of her own role. Many fervently believe in what they think they can do. Still, I suspect that some may be more knowledgeable about the truth than they let on. In one sad demonstration, a facilitator allegedly reported what an autistic child thinks of the test data: "I am extremely angry. . . . Please encourage us to become a part of your world and get the hell out of our world." As Brugger notes, the sincerity of the facilitator and the refusal of the facilitator to acknowledge her own causal role in the messages is a little reminiscent of automatic writing as seen in the case of Staudenmaier.

I do not know whether Brugger's analogy to pathological cases of movement confusion holds for facilitated communication or whether most of it is just everyday wishful thinking, but his hypothesis is worth noting.

ABERRATIONS IN SELF-CONCEPTION

MANY FORMS of self-deception can be found in the world of human behavior, and those we do not share make us scratch out heads. For example, a completely different kind of delusion, not related to movement ownership, is Cotard's syndrome, in which a subject may be convinced that some of his organs are missing or that his limbs are paralyzed or that he really is dead, appearances notwithstanding. In one case, the patient believed his stomach was missing and indeed lost more than 20 pounds in one month.

The delusions typical of Cotard's syndrome seem to be an expression of a rare psychiatric condition that may be part of a severe depression, such as in the depressive phase of a bipolar disorder or of some other psychotic condition. It may also occur, though rarely, after traumatic brain injury or as a consequence of neurosyphilis or multiple sclerosis. The patient is likely to have other neurological signs as well, such as refusal to eat or move or speak.[16] Successful treatment strategies include anti-depressants, electroconvulsive therapy, and a drug to enhance dopamine. Some cases do not respond well to any intervention.

An equally puzzling delusion is often seen in patients with advanced anorexia nervosa. They look at themselves in the mirror and report seeing their body as chubby—that they need to lose weight. This is puzzling, since her own image seen in the mirror is seen by her as distorted. Here is an interesting test: Show an anorexic patient a special narrow door, and ask whether her body can pass through that door and also whether a healthy control subject can pass through that same door. Regarding the healthy control, the anorexic patient says yes, but with regard to herself, she says no, she is too broad to pass through the door.[17] Still, one wonders: How different is that from those of us who persist in a thin body image while just overlooking a growing potbelly?

Delusions are generally very difficult to address neurobio-logically, as they may have widely distributed causal factors and there may be significant individual variation. Nevertheless, Bud Craig's framework does suggest one speculation regarding Cotard's delusion: a disruption of the pathways essential for integrating signals regarding the state of the body with evalu-ative and emotional signals. Following such a disruption, you might register the sensation that your foot hurts, but the pain itself is registered as neither a good thing nor a bad thing. You are indifferent. The feeling of mattering (the emotional valence) is detached from the report on the state of the body. Nothing matters anymore. It is like being dead. Is the insula messed up in Cotard's? Maybe, but it is complicated. Reports of patients with bilateral insula damage but preserved self-awareness suggest that self-awareness is multidimensional, implicating an interconnected set of regions, probably including, as Antonio Damasio has long argued, the brainstem.[18]

DECISIONS, UNCONSCIOUS PROCESSING, AND SELF-CONTROL

WE MAKE decisions every minute we are awake. Many are routine and are largely handled by our trusty habits, with just a bit of input from conscious perception and feelings. But sometimes the situation calls for focused attention if the perceptual input is ambiguous or unexpected (is the horse limping? is the chain loose on the chainsaw?). Sometimes we have to make momentous decisions, where we take more time, gather information, seek other opinions, and so forth. A single parent may be trying to decide whether she can keep her ailing, somewhat demented mother at home or whether it would be best for all if her mother were transferred to an extended care unit. Or a child is diagnosed with severe attention-deficit-hyperactivity disorder (ADHD) and has been suspended from school for being excessively disruptive. The parents have to decide whether he should be given methylphenidate (Ritalin). Or a woman is diagnosed with breast cancer and has to decide which of the treatment options to select. Each of these decisions is very difficult.

There is no algorithm—no exact recipe to follow—for good decision making. There are just very general bits of wisdom. Be reasonable, do your best, think it through, but don't dither. A decision indefinitely postponed while awaiting yet more data is sometimes catastrophic. A decision made quickly on the basis of "gut feelings" without evaluating evidence can be idiotic. All decisions have to be made in a timely manner, and the nonconscious brain tends to have a pretty good idea about when that is.

Once the relevant evidence is gathered and the future consequences predicted and weighed as best one can, the decision to be made is virtually always a decision under uncertainty. Sometimes the uncertainty is massive, but even then, some decisions are clearly worse than others. Feelings and emotions play a role, but not too much; other people's advice plays a role, but not too much. Decision making is a constraint satisfaction process, meaning that we assign values to certain things, subject to constraints. Thus, we may have many goals and desires about both the short term and the long term, and a good decision satisfies the most important ones.

For example, suppose I need to replace my old SUV because it is falling apart. I need a new car. What features do I want? Brand new, space for the dogs, safety, a rearview camera, leather seats, four-wheel drive, not too pricey, good gas mileage, reliability, and comfort. These goals may not be mutually satisfiable, I discover, as I go through the vehicle lists at Edmunds.com. Some may have to be downgraded to the status of "nice-idea-but." I may have to go up in price and down in leather seats or in four-wheel drive. Do I really need that? Or I may make other trade-offs. I may discover that a Japanese brand I thought was the most reliable is now equaled in reliability by a U.S. brand that is cheaper. And so forth. *Precisely* what my dear old brain is doing as I go through these exercises is not entirely known. That is, we can think of it in terms of constraint satisfaction, but we are still a bit vague about what constraint satisfaction really is in neural terms.[19]

Roughly speaking, we do know that in constraint satisfaction operations, the brain integrates skills, knowledge, memories, perceptions, and emotions and somehow, in a manner we do not precisely understand, comes to a single result. I can only buy one car. I can only go golfing or go swimming, not both at the same time. Somehow things are weighed and evaluated, and some evaluations can shift about as new data come in, while others may remain fixed. If we notice that we tend to make a certain kind of error in our past decisions—for example, biasing in favor of money over time or red flashy cars over higher-quality white cars—we can try to correct for that in the future.

The unconscious brain contributes much to the process of constraint satisfaction. Wise evaluation and intelligent decisions would be impossible without it. By learning and thinking and developing good habits, we give our unconscious brain better tools to work with. That is as much a part of developing self-control as sticking a reminder note on your computer that says, "Never respond to Evan's outrageous e-mail messages in less than 24 hours."

In neuroscience laboratories, decision making is being studied in monkeys and rats. One aim is to understand the fundamentals of how the neurons integrate signals from one sensory source over time (evidence accumulation).[20] Another goal is to understand how signals from different sensory sources, such as sound and touch, are integrated to enable a judgment.[21] With that in place, other aspects of integration can be addressed, such as how appropriate memories are recalled and integrated to help evaluate the significance of a sensory signal relative to the animal's goal. It is of course most productive to start where you can make progress, and that means starting with conditions that are not so complex that the data are uninterpretable. The research on integration for decision making, along with research on self-control and on the reward system, will increasingly be interwoven to give us a richer understanding of how we humans make decisions.[22]

Some economists have explored human decision-making behavior in tightly controlled experimental conditions, with revealing results. Called neuroeconomics, this research involves structured games, and the subjects play with real money.[23] In the game known as Ultimatum, one person (proposer) is given a sum of money, say $10, and he can propose an offer (between $0 and

$10) to another player (the responder). If the responder accepts the offer, they both keep the money—what is offered is kept by responder, and what is held back by the proposer is kept by him. If the responder rejects the offer, they both get nothing.

What do people do? First, notice that this game is played exactly once, between players who do not know each other and have no reason to see each other again. So, to be perfectly rational, according to economists, the responder should never reject any offer above $0. In other words, even a measly $1 is always better than $0. Nevertheless, many factors affect what responders do, indicating that there may be reasons why $1 is not always better than $0. For example, many responders feel insulted if the offer is less than $4 and would rather take nothing than an insulting offer. If you size this up in strictly monetary terms, the response may seem irrational—the proposer is not someone the responder knows or will know. Moreover, proposers seem to be aware of this likely rejection of very small offers, and they typically do not offer a sum they suspect with be rejected as unfair.

Why care about a very low offer? But we do. I do. I realize, as I think about it, that I would be insulted and reject an offer less than $4. I also know that as a proposer, I would offer $5, so as not to risk making the responder feel bad. I suspect that many economists would regard this as sheer silliness, but that is both my first *and* my considered response. My brain, in this instance, may be telling me—and the economists—something useful.

The thing is, most of my social interactions are not like these artificial games in the laboratory. They typically involve people I may be acquainted with and perhaps know well, who know people I know, with whom I will work with again in the future, and so on. So my habits of adhering to decency and fairness, acquired in childhood, will weigh against my accepting or offering what seems to be an insulting offer. Importantly, one major outcome of this neuroeconomic research has been the exploration of the role of emotions in good, intelligent decision making and a reconsideration of narrowly defining rationality in terms of maximizing monetary income.

Rationality also involves assessing the impact of my action on future interactions with kith and kin. If you have the reputation of accepting insulting offers, your respect among others will sink. That is a social cost—not one that can be evaluated in dollar terms, but a real cost. This may help explain some of the rejections of low offers. By and large, it is anything but irrational to keep these evaluations concerning reputation in mind. If I tarnish my reputation by being stingy or by being a wuss, I may pay dearly later. These calculations on my part may go on unconsciously, but they figure importantly in the decisions I make.

Aristotle understood, as did many later philosophers, including certainly David Hume and Adam Smith, that emotions play an important role in wise decision making. They knew that stylizing rationality as following a rule or principle that seems rational, though having a simple logical appeal, often leads to disaster. That is because no single rule can cover every contingency that can crop up in the world, and some contingencies are so surprising that they call for wise judgment, not slavishly adhering to a rule.[24]

Even the venerated Golden Rule—*do unto others as you would have them do unto you*—does not cover all contingencies. For example, as a scientologist, you might very badly want me to become a scientologist, on grounds that were you me, that would be what you would want. But I do not want to be a scientologist. So, say I, stop trying to help me become one. Moreover, some truly dreadful things have been done under the conviction that applying this rule was right. In the early part of the twentieth century, well-meaning Canadian bureaucrats removed Indian children from their homes and families in the bush and placed them in residential schools in cities like Winnipeg and Edmonton, far away from home, in expectation of integrating them into the wider white society. They thought that is what they themselves would have wanted had they been living in camps in the bush. The results were catastrophic.[25] This does not mean that the Golden Rule is useless as a *rule of thumb* or as a teaching tool. Nonetheless, it does mean that it cannot be assumed to be *the* fundamental moral rule that must always be followed.

Even the less intrusive version of the Golden Rule—*Do not do unto others what you would not want done unto you*—is beset with problems. It relies on the assumption that all of us pretty much share the same moral perspective, that we are all morally decent. Those who are genuinely evil can always invoke the rule to their own ends. Thus, the Nazi can say, "Well, yes, were I a Jew I would not want you to spare me the gas chamber." How can he say such a thing? Merely by being a consistent Nazi.

None of this means that we should always trust our feelings, disregarding all advice to carefully think things through. Propagandists, demagogues, and advertisers are all too ready to manipulate our passions for their own selfish ends. Reason certainly needs to play a role, but whatever reason is exactly, it also needs the balance of emotions to aid in the evaluation of possible consequences.[26]

Aristotle made the point that developing good habits as early as possible is a sound if not infallible guide to living well. With strong habits, we make good decisions even under strain and stress. And often those habits provide the unconscious compass concerning what is relevant and what is not, what is important and what is not, what is valuable and what is not, that allow us to navigate the social world.

Aristotle was a man of the world, and he addressed these issues intelligently, without referring to the gods or to an afterlife. He wisely advised that we should cultivate habits of moderation in all things, the better to dodge the avoidable tragedies of life. He advised us to be courageous, but neither reckless nor cowardly; thrifty, but neither miserly nor extravagant; generous, but neither too much nor too little; persistent, but knowing when to change course; tending to our health without being either excessively indulgent or excessively preoccupied. And, sensibly, so forth. Aristotle is not a charismatic figure, as perhaps some televangelists or talk show hosts may be. He did not profess to have a special secret to happiness and the well-lived life; he did not require special rituals or observances or costumes or hats. But then he was not trying to sell anything. He was merely making careful observations about what in his considered judgment is likely to serve you well in living your life. As I look at things now, his advice, while not dazzling, seems profound. The dazzling, by contrast, may conceal creepy motives or hunger for power.

THE MORE I learned about nonconscious processes in the brain and how they seamlessly interweave with conscious processes, the more I began to realize that to understand consciousness, we need to understand more about those nonconscious processes.[27] When we embrace what we learn about nonconscious processes, whether in perception, decision making, or casual social behavior, we expand our self-conception. We understand ourselves more deeply. I think that Freud (the later) was off the mark when he thought we learn a lot about our unconscious through slips of the tongue or free association or dream analysis. I am not sure we learn anything much at all using those strategies. By contrast, psychologists studying visual and other illusions, neurologists studying human subjects with brain damage, neuroscientists studying the reward system, and neuroeconomists studying decision making teach us a lot about ourselves. They have discovered things we could not have known merely by introspection or dream analysis.[28]

So what about consciousness? What sort of brain business *is* it? What do we know and what can we know? That is the topic of the next chapter.

Chapter 9

The Conscious Life Examined

THE NEUROBIOLOGY OF CONSCIOUSNESS can be addressed through different strategies, each of which targets this question: What are the differences in the brain between conditions when we are conscious and those when we are not? Once those differences begin to be clarified, then the next question is this: What mechanisms support and regulate conscious states? With progress on those two questions, we may be able to address why hunger and thirst, or sound and sight, or the passage of time and the relations in space are experienced in the unique ways that they are. The neurobiology of consciousness is not a single problem in the way that the structure of oxytocin receptors, for example, is a single problem. It is a many-factored problem.

We can begin with something we all know about: the loss of consciousness during deep sleep.

SLEEP AND THE LOSS OF CONSCIOUS EXPERIENCE

EVERYONE SLEEPS. We may sleep different amounts, we may be on different schedules; some are larks, others are owls. Some people are easier to arouse from sleep than others. But everyone sleeps. About a third of our lives is spent in sleep. But not because we are lazy.

The most basic experience we all have of the difference between being conscious and not being conscious is falling asleep. In the state of deep sleep, we are essentially unaware and without conscious experience. Nor do we engage in purposeful action, such as picking apples. Our goals and plans are on hold. We have no memory of being in deep sleep. It is almost as though the conscious self ceases to exist, like a fire quenched.

Consequently, sleep has long been a compelling entry point for asking the fundamental question about consciousness: What is different about the brain when we are conscious and when we are not? If we can make progress on the distinct neurobiological profiles of being in deep sleep and being awake or dreaming, it may help us to answer this: How does the brain generate conscious experiences?[1]

Sleep seems to be necessary for all mammals and birds, possibly all vertebrates. Surprisingly, it is necessary even for some invertebrates. As Ralph Greenspan demonstrated, fruit flies sleep. Moreover, caffeine keeps fruit flies awake, and they respond to anesthetics such as ether much as humans do. How do they behave when they sleep? They sit very still on their perch, not unlike certain birds. Does this mean that when they are awake fruit flies are *conscious*? No one knows exactly how to answer that yet, but it is worth pondering. In any case, the discovery that fruit flies sleep points to how remarkably ancient, evolutionarily speaking, sleep is.

Moreover, sleep is essential. Animals deprived of sleep will make it up when allowed to do so, a phenomenon known as *sleep rebound*. Fruit flies deprived of sleep also display sleep rebound. Rodent studies show that rats deprived of sleep but otherwise well tended fairly quickly show a range of abnormalities, including weight loss, body temperature changes, and immune system changes. With continued sleep deprivation, they die within weeks.[2] This is probably true of humans as well.[3]

College students regularly entertain the conviction that they could be more efficient and get more work done if they slept less than they normally do. Sleep is a waste of time, they suppose. This is completely false. Your brain is not doing nothing while you are asleep. One of its jobs during deep sleep is consolidation of memory—transferring and organizing important information acquired during the day to long-term storage in cortex while culling out the unimportant stuff. So students who force themselves to sleep less are likely depriving their brains of the very thing they need for the grades they seek: consolidation of memory.[4]

Despite our efforts to cut back on sleep, each person's brain has a set point for the amount of sleep needed, which varies only a little between individuals of the same species.[5] Legendary nonsleepers such as Thomas Edison and Winston Churchill did in fact sleep—they just napped a lot. Eventually, alarm clocks and caffeine are no match for the biological imperative to go to sleep. After a brief spate of trying to cut back on sleep time and observing that our efficiency while awake suffers, we all resume sleeping the amount we need. But the evolutionarily ancient character of sleep implies that it is important for even more deeply biological reasons that we have not yet uncovered. That is, even if memory consolidation operates during human sleep, it is doubtful that memory consolidation is the main function of sleep in fruit flies. However sleep benefits fruit flies, it probably benefits us in much the same way.[6]

SLEEP MY CHILD AND PEACE ATTEND THEE, ALL THROUGH THE NIGHT

THIS FIRST LINE from the beautiful Welsh lullaby asks that peace attend, for indeed, a peaceful night of sleep is a blessing; a disturbed sleep, a misery. There are various kinds of sleep disorders, and some are especially significant as neuroscientists try to figure out the underlying neural basis of deep sleep and dreaming and how these states differ from each other and from the waking state.[7] Let's first consider sleepwalking.

Scotty walks in his sleep. This is not a stunt. Sleepwalking is a dangerous and profoundly biological brain disorder. Scotty is bright, engaging, well loved, and well brought up. He has lots of friends and did splendidly in college. Now in his early 20s, he suffers sleepwalking episodes several times a month. Still worrisome, this is at least an improvement over the nightly somnambulisms of his childhood. At age 3, he would begin to scream about 1.5 hours after being tucked in, causing his worried parents to sprint to his bedside. If his bedroom door was not locked, he might, in a state of attenuated awareness of his actions and intentions, walk downstairs, out the front door, and on down to the barn. Or he might raid the fridge or curl up on the back porch. Why? He has no awareness of why or how.

At 8 years old, Scotty awoke to find himself sitting on a bridge over the creek at the back end of the farm. The cold finally woke him up, and he trudged home, frightened and upset. All this, after he had carefully tied one end of a rope to his foot and the other to the bedpost as he settled in for the night. To no avail: come morning, the rope lay unknotted on the floor by the bed.

Once he walked out of the house and opened the gate to the pigs' pen. By morning, five pigs were nowhere to be seen, and much of the rest of the day was spent rounding up freedom-loving, garden-gobbling pigs. Seeking ways to outfox his sleeping brain's proclivities, he contrived a spray bottle that would spray his face when he pushed down on the handle of his bedroom door. That sometimes worked to jolt him awake; sometimes it just got dismantled.

Scotty never has any memory whatsoever of the events during his sleepwalking. With the morning light, he will discover that he has freed the pigs, for example, but his memory for his nightly shenanigans is completely and utterly blank. He is sad and embarrassed. If Scotty has a bedtime wish, it is that he will not somnambulate.

What is Scotty's sleepwalking state like? What is he aware of? Because Scotty has no recollection, he, too, wonders about that very question. Evidently, in his sleepwalking, Scotty's brain has sufficient sensitivity to his surroundings to navigate paths, remove tethers, and open—and even *unlock*—doors. His eyes are open but he does not seem to have a normal visual take on his surroundings. He sees a door, but not his parents standing by. He seems to be aware of, or at least acutely responsive to, his feelings of anger and will sometimes call out "Get it, get it!" with shrill intensity. But his consciousness during these periods is nonstandard, to say the least. Scotty does not respond to queries or commands, such as "Get what?" Yet once when his sleepy and exasperated brother clapped his hands and yelled, "I got it!" Scotty instantly relaxed and went into a sound sleep. When sleepwalking, he is not aware of the inappropriateness of certain actions, such as wandering the farm in his underwear and bare feet. He would be agonizingly aware in his normal awake state of just how bizarre his behavior is. This lack of self-awareness is consistent with being asleep, not with being awake.[8]

Sleep scientists believe that in sleepwalking disorders (*parasomnias*), some form of waking intrudes into deep sleep. The two brain states—being in deep sleep and being awake—are radically different in most humans. When you are asleep and not dreaming, consciousness appears to have shut down. You do not make purposeful movements, except perhaps rolling over in bed. You are aware of nothing, though you can be aroused by strong stimuli, such as a loud noise. When you are awake and alert, you are aware of what is going on around you, of motives and intentions and thoughts. These differences between deep sleep and the waking state are reflected in the recordings (electroencephalograph, or EEG) taken when electrodes are placed on the scalp. In deep sleep, the EEG waves are long and slow. In the awake state, they are sharp and fast. What exactly these differences mean in terms of the activity of the neurons under the scalp is, at this stage, only partially understood. The basic point is that the waveforms reflect an aggregate of the activity of many millions of neurons.

Is Scotty just acting out his dreams? No, as it turns out. The evidence for this judgment rests on several observations. First, a special cap fitted out with electrodes capable of recording large-scale brain events shows that Scotty's sleepwalking does not occur during dreaming (Figure 9.1). How do we know that?

 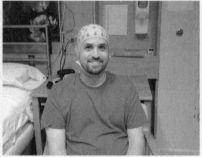

9.1 *The photo on the left shows a child sleeping with EEG electrodes on her face and scalp as her sleep stages are recorded. The photo on the right shows an EEG cap with the electrodes already fixed in place, which is a little more convenient than attaching the electrodes directly to the scalp.* Left photo courtesy of Rebecca Spencer, University of Massachusetts Amherst. Right photo courtesy of Neal Dach, Harvard University.

In the mid-twentieth century, sleep scientists noticed an interesting correlation: During certain times when you are asleep, your closed eyes will rapidly move back and forth. If, during those periods, you are awakened, you will likely report that you have been having visually vivid dreams. Further study showed that *rapid eye movements* (REM) do not occur during deep sleep, which scientists therefore referred to as non-REM sleep. For a time, sleep scientists suspected that rapid eye movements were a behavioral signature that you were probably dreaming. That would have been convenient indeed, but more penetrating studies revealed that subjects awakened during non-REM sleep are frequently apt to report mentation of some kind, though not typically as visually vivid as during REM sleep. Subjects awakened during non-REM sleep might say that they were thinking about writing an exam, for example. Or they might say that they were aware of nothing. (See Figure 9.2 for the sleep stages and the transitions between them.)

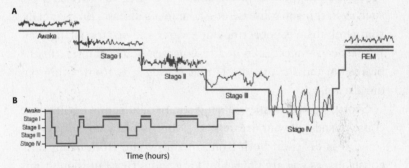

9.2   *The different stages of sleep. (A) The distinct patterns of activity recorded using EEG during the four stages of sleep and during the awake state. Stage IV is also known as deep sleep. REM refers to rapid eye movements, which are characteristic of dreaming sleep. REM periods are indicated by the black bar. (B) The approximate time spent in each of the stages at different points during the 8-hour sleeping period. Notice that deep sleep tends to occur only very early in the sleep cycle, whereas dreaming (REM) is infrequent early in the night but increases as the night goes on. There is much variation between individuals and within individuals from birth to old age.* Adapted from Edward F. Pace-Schott and J. Allan Hobson, "The Neurobiology of Sleep: Genetics, Cellular Physiology and Subcortical Networks," *Nature Reviews Neuroscience* 3 (2002): 591–605. Copyright © 2002, rights managed by Nature Publishing Group. With permission.

Recordings from Scotty's brain during his sleepwalking episodes indicate that he is not dreaming. Incidentally, the brain
waves recorded during REM are more similar to those recorded
during the waking state than to those recorded during deep sleep.
Still, dreaming is not just like being awake, since your dream content can be bizarre without your recognizing that it *is* bizarre.
Sheep fly, bears talk, long-dead great grandfathers climb trees,
and you are not surprised. One known contrast in brain activity
between being awake and dreaming is that the prefrontal cortex
(PFC) is much less active during dreams than in the awake state.
This lowered activity correlates with a dulling of your critical fac-
ulties and hence the lack of surprise when bears talk.

Although Scotty is not acting out his dreams, there are indeed
sleep disorders in which sleepers do act out their dreams. This
sometimes happens in older adults with degenerative diseases
such as Parkinson's disease or Alzheimer's disease. But typically,
somnambulism is not acting out dreams and occurs during non-
REM sleep stages. As noted in Chapter 3, mechanisms in the
brainstem induce a paralysis that prevents us from acting out
our dreams.

Scotty's consciousness during somnambulism is not like the
consciousness of your dream state. Nor is it like your fully awake
state. It is different both from dreaming and from being fully
conscious, as you are now. Thinking about that raises questions
about what the brain *is* doing when we are fully conscious and
what it is doing when we are *not*. There has to be a difference.
These contrasts are helpful in moving forward toward understanding the brain basis for consciousness. We can then ask:
What is the nature of the differences? In what ways does the
dreaming state contrast with both deep sleep and being awake?

The neuroscientific research addressing these questions is
still in its early phase, but inching along on a range of fronts,
scientists have made much more progress than many nay-saying
philosophers thought possible.

THE SCIENCE OF CONSCIOUSNESS

WHAT DO WE KNOW? First, there is no circumscribed brain location—no anatomically discrete module—that is the seat of conscious experience. There is no single place you can point to and say, consciousness is exactly *there*.[9] But then that is true of memory, self-control, action organization, and just about everything else. Nor is consciousness the exclusive job of one hemisphere, but not the other. Second, not all parts of the brain are necessary for conscious experience. You could lose a sizable chunk of cortex in the region above your left eye, for example, but you would still have normal conscious sensory experience. If you then also lost a matching chunk over the right eye, you would still have largely normal sensory experience, though your emotional responses would likely be altered. You might, for example, be less likely to feel disturbed by the sight of a street brawl or less likely to feel socially awkward when a dinner guest refuses to eat your food. Very roughly, such is the state of people who underwent a prefrontal lobotomy, a surgical procedure disconnecting the prefrontal cortex from other structures and used extensively for a range of psychiatric conditions from about 1943 to 1955.

On the other hand, there are particular structures in the brain, along with the looping links connecting those structures, that are necessary for maintaining consciousness. Exploring these structures and their links has turned out to be very fruitful.

As a preliminary, note that it is useful to distinguish between the structures that support being conscious of *anything at all* and structures that contribute to being conscious of this or that—the so-called *contents of consciousness*. If you are in coma, for example, you are typically not aware of seeing or hearing or smelling *anything*. When you are awake, you are aware, for example, of the sight of a dog, his bark, and his doggy smell.

Those are the so-called contents of consciousness: awareness of specific events.

Nicholas Schiff is a neurologist studying disorders of consciousness. He wants to figure out what goes wrong when we cease to be aware of anything at all. His explorations have led him to the central thalamus and its incoming and outgoing pathways.[10] According to Schiff's hypothesis, to be conscious of anything requires activity from a ribbon of neurons in the middle of the thalamus, whose activity is itself regulated by neurons in the brainstem, an evolutionarily very old structure (Figure 9.3). Called the *central* thalamus (also called the *intralaminarnuclei* of the thalamus), its neurons have pathways, though sparse, to the top layer of *every* part of the cortex. That organization is unique and suggests that consciousness involves the upregulation of the entire cortex, whereas the reverse, loss of consciousness, is related to downregulation. In both cases, the changes are dependent on the activity of neurons in the central thalamus.

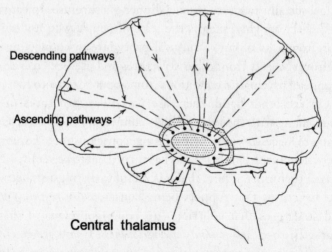

9.3 *Highly schematic characterization of the central thalamus (a ring of linked nuclei within the thalamus, here shown as dotted) and the projection pattern to the upper layers of the cortex.* Courtesy Paul Churchland.

Here is another striking fact about neurons in the central thalamus. There are looping neurons from these upper layers of cortex projecting right back to the ribbon in the central thalamus. The looping back allows for maintaining an especially potent but transient connection for a chunk of time, as, for example, while paying attention to a particular sensory event or feeling.

To see why the projection pattern of the central thalamus seems a big clue to consciousness, consider the contrast with the organization of those regions of the thalamus that relay information from the sense organs. For example, neurons from the retina connect to a specific region of the thalamus (the lateral geniculate nucleus, or LGN), but the LGN then projects *only* to visual cortical area V1—not to everywhere, not even to everywhere in the visual cortex (Figure 9.4). Analogously, neurons in the cochlea that are sensitive to sound project *only* to auditory cortical area A1. And so on for other sensory systems. This pattern of processing suggests a system-by-system development of a specific signal, whereas the pattern of the central thalamus suggests a different set of functions: be-awake-and-alert or downregulate-and-doze. This separation of structures and functions in the thalamus seems to reflect the main differences between being aware of anything and being aware of something concrete and specific.

Another distinction that demarcates the roles of the central thalamus from those of the other sensory-specific thalamocortical systems is the style of neuronal activity. Those central thalamus neurons with their unique connectivity also have unique behavior. During the awake and dreaming states, neurons in the central thalamus fire in bursts at an unusually high rate—800 to 1,000 times per second (hertz), a remarkably energy-intensive behavior not seen anywhere else in the nervous system. They do not display the bursting pattern during deep sleep.[11] The neuronal bursts of central thalamus neurons track the aggregate brain wave pattern seen on the EEG that typifies the awake and dreaming states—20 to 40 hertz.

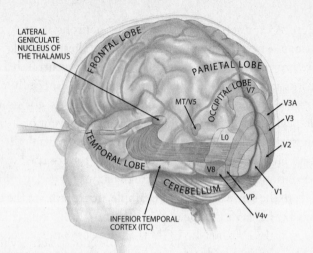

9.4 *Pathways from the retina to the optic chiasm, where the pathways cross, to the lateral geniculate of the thalamus. From the thalamus many fibers travel to the first visual area of cortex, V1. Auditory signals, not shown, go from the ear to the medial geniculate of the thalamus, and then to the first auditory area of the cortex, A1. A similar pathway pattern is also seen for touch and for taste.* © 1999 Terese Winslow, with permission.

Collectively, here is what the clues suggest: the ribbon of neurons that is the central thalamus is controlled by activity in the brainstem and in turn regulates the cortical neurons to ready themselves for conscious business. Activity in this three-part arrangement—brainstem + central thalamus + cortex—is the support structure for being conscious of anything at all.

Damage to the central thalamus has serious consequences for consciousness. If a lesion occurs on one side of the central thalamus, then the person tends not to be conscious of or attend to events related to the affected side (which is always on the opposite side of the body, because in the brain there is a cross-over of pathways—right brain controls left side and vice versa). What happens if there is damage on *both* sides? The person is in coma. This was the first clue that the central thalamus has a special role in consciousness. If the lesion is small, the coma is likely transient. How could that be? As long as other neurons in the central thalamus are healthy, they probably sprout a little and take over the functionality of those that died. Let's look now at a remarkable patient studied by Nicholas Schiff.

Schiff had a patient, Don Herbert, who had been in coma for nine years. He had been a fireman who suffered oxygen deprivation when the roof collapsed in a building where the team was fighting a fire. Although he briefly regained consciousness, he then lapsed into coma, a state where he remained for nine long years. His brain scan indicated a normal-looking cortex, with no visible lesions, holes, or atrophy. A scan that is sensitive to neuronal activity—a PET scan—showed reduced activity in the central thalamus. Ever hopeful, his family included him in their events, though his head lolled while he sat in his wheelchair, unaware of events around him. Then one day, his physician changed his daily cocktail of drugs to keep him going and included also the drug Ambien. Yes, the same drug that can induce sleep. To everyone's astonishment, a few weeks later Herbert woke up. Confused as to his whereabouts, but using language, he asked what was going on. Whether the Ambien was important in his waking seems plausible, but nevertheless, the causal relationship is not known for sure.

How should we understand this astonishing and abrupt recovery? Based on his knowledge of the physiology of the brain, Schiff suggests that following the loss of oxygen during the fire, the ribbon of neurons in the central thalamus had ceased to function normally, so their normal activity supporting waking and arousal was shut down. It was as though Herbert was in a state of perpetual deep sleep—unarousable deep sleep. Conceivably, the Ambien might have kick-started the dormant sleep/wake cycle, and as a consequence, Herbert awoke; he regained the capacity to be conscious of things going on around him. He saw, heard, and moved his arms appropriately. He recognized family members, though his son had grown up, and he remembered events from his past. Had the cortex itself been severely damaged, his recovery likely would have been impossible. But with a healthy cortex more or less waiting to transition into an awake/consciousness mode, once the central ribbon of neurons were bursting and back in vigilance state, so was awareness of sights and sounds and touches.[12]

It should be noted that most patients in chronic coma or in persistent vegetative state are unlike Mr. Herbert. More often than not, they suffer severe damage to large areas of the cortex, damage that can be seen in a brain scan.

These results also take us back to thinking about Scotty and his sleepwalking. Might it be that during his sleepwalking episodes, some part of his central thalamus is in awake mode, but other parts, such as that section of the central thalamus that projects to the prefrontal cortex, are not? Suppose that during sleepwalking there is enough activity in the sensory cortices to guide action, but not enough in the prefrontal range to yield full alertness and allow for memory. As we shall see in the next section, this is not implausible.

One further avenue of research should be mentioned first, and that concerns epilepsy. There are many types of epilepsy, but the one that may present a special opportunity for exploring the mechanisms of consciousness is a type known as absence sei-zures. These seizures are most common in children and consist of a temporary cessation of activity, lasting about 10 seconds. During this period, it is as though the child's mind has gone somewhere else; no one is home. Hence the name *absence seizure*. During the seizure, the child may show some mild twitching or finger tapping, but not much else. Talking abruptly ceases. When the child resumes normal activity, she will have no memory of the lapse and carries on as though nothing has happened. The severity of the seizure varies across subjects. If a person's brain is imaged during an absence seizure, the most consistent finding is a decrease of brain activity in some prefrontal regions, par-ticularly the prefrontal areas and their connection to subcortical structures, and an increase in others.[13]

LINKING CONSCIOUSNESS OF *SOMETHING OR OTHER* TO CONSCIOUSNESS OF *SOMETHING IN PARTICULAR*

TO BE AWARE of a baby crying or the sight of Mount Baker, the system involving the brainstem, central thalamus, and upper layer of cortex has to be in its *on* state. Central thalamic neurons must be firing in bursts that ride the lower frequency brain wave of about 40 hertz. In addition, the specific areas of the thalamus (for sound and sight, respectively) must be talking to proprietary areas of the cortex. That is the hypothesis.

Suppose you are paddling a canoe down a river, looking out onto the water, sensitive to the current and to what might be beyond the next bend. The scene is complex, and you observe it in spatial depth, suddenly realizing that there is a bull moose peering at you from the bank, between the chokecherry bushes. You have this thought: this is mating season, and the bull is apt to be aggressive. You are wary and steer a wide berth. You listen also for changes in river sounds that signal changes in current and for calls that might signal other moose nearby. None of this is automatic, and all of it requires vigilance, exercise of skills, and retrieval of knowledge from memory.

The underlying processes that lead to recognition of a bull moose are not available to consciousness, nor are the processes that retrieve from memory the information about moose in the fall rut, nor are the processes that underlie vigilance and shifts of attention. That is all beneath your awareness. It is your silent thought, your dark-energy thought. But your conscious awareness of the scene and your focused attention to the movement of the water and the behavior of moose are necessary as well. You could not do this in your sleep. How does all this work?

In about 1989, psychologist Bernard Baars proposed a framework for research on consciousness with a view to fostering a coevolution of psychology and neurobiology. Baars realized that to understand how we are conscious of specific events, we will need to know a lot about the entire brain, the nature of the pathways between neuronal pools, and how subcortical structures interact with the cortex. He delineated the problem by first listing the significant psychological properties and capacities associated with being aware of specific events in consciousness.[14]

First, he emphasized that sensory signals of which you are conscious are highly integrated and highly processed by lower-level (nonconscious) brain networks. That is, when you hear the captain explaining that the plane will be delayed by fog, you are not first conscious of a string of sounds, then conscious of figuring out how to chunk the string into words, then conscious of figuring out what the words means, then conscious of putting it all together to understand the meaning of the sentence. You hear just the captain; you are aware of what he meant.

Second, the information stored concerning the behavior of moose in the fall and about changes in river currents are suddenly consciously available to help you decide what to do in this novel situation. This means there must be integration of sensory signals with relevant background knowledge—with *stored* information.

The third important point is that consciousness has a limited capacity. You cannot follow two conversations at once, you cannot at the same time do mental long division and watch for dangerous eddies in a fast-moving river. When we think we are multitasking, we are probably shifting attention back and forth between two or possibly three tasks, each of which is familiar and which we can perform with minor vigilance.

Fourth, *novelty* in a situation calls for consciousness and for conscious attention. If you are fighting a barn fire, you must be alert and vigilant. On the other hand, if you are a veteran cow milker, you can milk the cow and can pay attention to something else.

Fifth, information that is conscious can be accessed by many other brain functions, such as planning, deciding, and acting. The information can be accessed by the speech areas so that you can talk about it. Conscious information is kept "on the front burner," so to speak. That is, the information is available for some minutes in working memory so that your decisions are coherent and flow sensibly together. The widespread availability of a conscious event was a hypothesis that Baars proposed, not an established fact, but it seemed completely plausible and provoked other questions, such as the regulation of access and the range of functions that can have access.

None of these five features is a blockbuster on its own, but notice that collectively they yield a sensible and rather powerful framework for guiding research into further matters, such as how information is integrated and rendered coherent in our experience. Wisely, Baars avoided trying to identify the *essence* of consciousness, realizing that essences are an old-fashioned way of thinking about phenomena that impede making actual progress. This contrasts with the approach favored by some philosophers, whereby they tried to identify the defining property of consciousness, such as self-referentially, which is knowing that you know that you are feeling an itch or pain.[15]

Baars called his framework the *global workspace theory*, using the metaphor of a workspace to convey the rich integration of information from different sources that characterizes conscious states, along with the availability to different functions of information in the workspace. *Workspace* draws on the idea that consciousness is a consumer of products of lower-level processing, that other functions such as decision making and planning can access what is in consciousness. Hence, the workspace is global.

The literal version of a global workspace looks like this: Suppose you work at Apple, and today you have a meeting about progress on the new MacBook Air under development. Each team sends a leader who describes their achievement—a tiny magnet plug, a solid state component, a superflat case, a slicker dongle for Ethernet, and so forth. At the meeting, the engineering details behind each achievement need not be gone into, just the product—the outcome of the engineering at the workbench. The interactions between the team leaders lead to upgrades of the products and plans for other projects. The meeting is brief, the next stage is arranged, and the team leaders go back to their particular work areas. The next meeting is about the new iPhone, and the integrative, executive process begins anew.

So assuming that you are conscious of the highly integrated results of processing, what about the unconscious processing that leads up to it? What is the difference in the brain between nonconscious processes and conscious processes? According to one hypothesis, a visual signal—say, the printed word *dog*—is first processed nonconsciously in visual areas and then becomes consciously perceived as the word *dog* only when the signal reaches areas in the more anterior cortical areas (temporal, parietal, and frontal lobes; see Figure 9.4). Likewise, an auditory signal is not consciously perceived unless the more anterior areas of the brain respond to that signal.

What is the evidence for this hypothesis? Experiments devised by Stanislas Dehaene and colleagues used an old technique called *masking* and put it to a new use. Here is how masking works. If the word *dog* is briefly flashed on your computer screen followed by a brief delay (about 500 milliseconds) and then by XXXXX, you will see (consciously) first *dog* and then XXXXX. However, if *dog* is followed *immediately* by XXXXX, you will not see *dog*. You will see only XXXXX. The word *dog* is *masked*. (We do not exactly know why, but it is.) The experimental idea of several labs was to scan the brain's activity both during the masked condition *and* during the visible condition in order to get some purchase on the critical question: What is the difference in the brain when you are conscious of the visual signal *dog* and when you are not?[16]

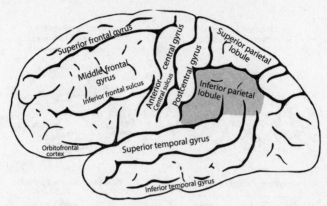

9.5 *Lateral aspect of the human brain, showing subregions of the parietal lobe, temporal lobe, and frontal lobe.* Adapted from *Gray's Anatomy*, public domain. Originally printed in Churchland, Patricia S. *Braintrust*. Princeton University Press. Reprinted by permission of Princeton University Press.

Before answering the question, I should mention that masking is not the only paradigm for addressing the questions concerning consciousness of perception and the brain. There are other visual setups, as well as analogous setups in other modalities. Here, however, I shall confine the discussion to visual masking, noting that there is impressive convergence of results across other paradigms and modalities.

So what is the difference in the brain between conscious processing and unconscious processing of a stimulus as revealed by the masking technique and brain imaging? Here is the simplified answer: When the visual signal *dog* is masked (not consciously perceived), only early visual areas (in the back of the brain) show activation. By contrast, when the visual signal is *consciously* seen, the posterior activity has spread to more frontal regions, including parietal, temporal, and prefrontal areas (Figure 9.6). Dehaene and Changeux refer to this as *global ignition*. This pattern of posterior to anterior spreading of activity seems to uphold Baars's hunch that conscious perception involves global connectivity in the brain, whereas unconscious perception is restricted to a smaller region.

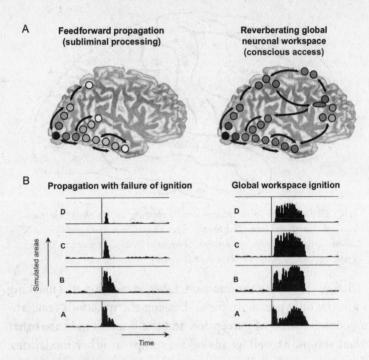

9.6 *Schematic representation of the events leading to conscious access. (A) Schema illustrating the main differences between subliminal and conscious processing. During feedforward propagation, sensory inputs progress through a hierarchy of sensory areas in a feedforward manner. Multiple signals converge to support each other's interpretation in higher-level cortical areas. Higher areas feed back onto lower-level sensory representations, favoring a convergence toward a single coherent representation compatible with current goals. Such a system exhibits a dynamic threshold: if the incoming activity carries sufficient weight, it leads to the ignition of a self-supporting, reverberating, temporary, metastable, and distributed cell assembly that represents the current conscious contents and broadcasts it to virtually all distant sites. (B) Simulation of two single trials in which an identical pulse of brief stimulation was applied to sensory inputs. Fluctuations in ongoing activity prevented ignition in the left diagram, resulting in a purely feedforward propagation dying out in higher-level areas. In the right diagram, the same stimulus crossed the threshold for ignition, resulting in self-amplification, a global state of activation.* Adapted from Stanislas Dehaene and Jean-Pierre Changeux, "Experimental and Theoretical Approaches to Conscious Processing," *Neuron* 70 (2011): 200–227. With permission from Cell Press.

The data from brain scans do not give us very accurate *timing* data concerning precisely when an event occurs. But a different recording technique does. Using EEG, which records aggregate brain activity by putting electrodes on the scalp, something interesting in the time domain emerges. A robust signature of conscious perception—a waveform in the scalp recordings—occurs about 300 milliseconds after the stimulus is presented. Significantly, these findings appear to hold also for other modalities.

This result suggests that the back-to-front global ignition required to yield awareness of a sensory stimulus takes about 300 milliseconds—about one-third of a second. This makes sense, since it takes time for a signal to travel from one neuron to the next, and a lot of territory has to be covered if you are a signal traveling from the retina to the prefrontal cortex.

One important development concerning the connectivity undergirding global ignition is the discovery that brains of all mammals, and possibly all animals, appear to have a *small-world* organization. Not every neuron is connected to every other neuron. If that were the case, our heads would have to be massively larger than they are. As neuroscientist Olaf Sporns and his colleagues have shown, any given neuron is only a few connections away from most other neurons. Brains have a "six degrees of separation" organization.[17]

Some neurons, like some people, are especially well connected. Thus, my friend Eric is a neuroscientist who is very well connected to media figures in New York, who in turn are connected to journalists in Washington, D.C. So through Eric I can connect to Anderson Cooper, for example, in only three or four steps. Some groups of neurons are hubs for putting other less well-connected neurons into contact with each other. The term coined for the especially well-connected neurons is *rich club* neurons. So a signal might follow this sequence: from a *locally connected* neuron to a *feeder* neuron, then to a *rich club* neuron, to a different *feeder* neuron, and then to a new *local* neuron (Figure 9.7).

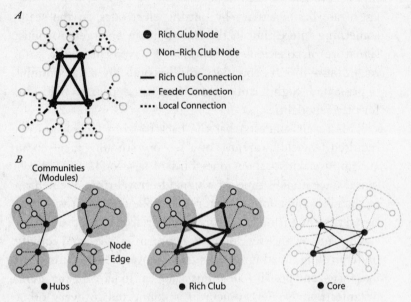

9.7 *(A) The relationships between rich club neurons and non–rich club neurons. Neurons in the local interconnected populations can access information from other local populations by going through the rich club routes. (B) Network communities (modules) consist of groups of densely interconnected neurons. The existence of several communities is characteristic of modular networks. Within-module connections tend to be shorter than between-module connections. In this way, spatial modules help to conserve costs related to wiring and communication, and improve the local efficiency of specialized neural computations. Functional integration between modules requires the addition of high-cost long-distance axonal projections to interconnect spatially remote brain regions. This gives rise to connector hubs, which receive a disproportionate number of long-distance, intermodular connections, have a high participation index, and occupy a topologically more central or "potential bottleneck" role in the network.* Part (A) adapted from Martijn P. van den Heuvel, René S. Kahn, Joaquín Goñi, and Olaf Sporns, "High-Cost High-Capacity Backbone for Global Brain Communication," *Proceedings of the National Academy of Sciences* 109, no. 28 (2012): 11372–77. With permission. Part (B) adapted from Ed Bullmore and Olaf Sporns, "The Economy of Brain Network Organization," *Nature Reviews Neuroscience* 13 (2012): 336–49. With permission.

The rich club organization is very efficient, which means that the brain can keep wiring costs down and head size can be kept within a reasonable volume (Figure 9.6). Wiring uses energy and takes up space, so a small-world organization with hubs is more efficient that having every neuron connected to every other neuron. And transient connectivity via the rich clubs can be very fast, brief, and efficient.[18] Of course, needs, drives, goals, and other internal signals will also play a role in whether you are conscious of a particular thing and hence whether the link followed is through *this* rich club rather than *that* one. Some events that are important in one context can be totally ignored in another.

Does the rich club framework help us connect the research on the central thalamus (how we can be conscious of *anything*) with the global workspace framework (what *particularly* we are conscious of)? Yes indeed. Consciousness of sensory signals involves linking spatially quite separate regions of the brain, regions that may be tightly coupled on a short-term basis. As attention shifts, the prevailing linkage weakens and other neuronal pools now take their turn in making strong links. The linkages, it is thought, may consist in synchrony in the activities of populations of neurons. So when your attention shifts from listening to a lecture on gut microbes to worrying about whether you have enough time to get to the bank, activity shifts from auditory cortex to areas along the midline of the brain, then shifts back again as you attend to the lecture.

So the current favored hypothesis is that the activity in neuronal groups is transiently synchronized, providing a coherent and conscious experience of, say, the gist of the lecture. The synchrony might be facilitated by the bursts of spikes displayed by the neurons in the central thalamus. The local connectivity of neurons within a group provides what we think of as context: your knowledge of the many kinds of microbes and their role in digestion. How exactly synchrony of activity achieves these effects is not understood, but researchers are hot on the trail.

Studies on how anesthetics work provide significant support for this hypothesis. Although much remains unknown about the details of the mechanisms of anesthetics, we do know that the general effect is to decrease the integrative chatter between neurons. Anesthetics change the lines of communication: some hubs are not answering the call just now; they are snoozing. The major effect of the anesthetic is to mess up communication between neurons. First, that means a disruption in the global ignition that facilitates the propagation of a sensory signal from the back of the cortex to more anterior regions of the cortex, such as the prefrontal areas.[19] Second, it means a disruption in the efficient communicative links between rich club neurons. In short, "anesthetics seem to cause unconsciousness when they block the brain's ability to integrate information."[20]

Accordingly, there are three properties that seem especially prominent in the neurobiology of consciousness: (1) rich club neurons and their ability to make fast connections to other rich club neurons, thereby providing the scaffolding for rich integration of information; (2) global ignition for brain events that reach consciousness; and (3) the central thalamus, with its role in enabling specific contents of awareness during the awake and the dreaming states. These three properties of brains suggest a platform that will certainly lead to a whole new range of experiments. With luck, these advances will eventually lead to understanding the detailed nature of the mechanisms involved in consciousness.

A COMMENT CONCERNING CONSCIOUSNESS IN OTHER MAMMALS AND BIRDS

BRAIN ANATOMY is highly conserved across mammalian species. The basic organization that is hypothesized to support consciousness in humans has been remarkably conserved during evolution of the mammalian brain and thus is remarkably similar across all mammals. The brainstem structures are highly similar, the thalamic organization is highly similar, the structures that regulate emotions are highly similar, as is the cortical organization. In short, the main elements supporting consciousness of anything are highly similar across all mammalian species. Size of the cortex differs across species, but there is no obvious difference in the cortex per se that suggests to neuroscientists that only humans have consciousness. On the contrary, it seems wholly plausible that consciousness, *in some form or other*, is a feature of the brain of all mammals and birds.

To be sure, small differences in the numbers of neurons or the size of a structure such as the cortex may yield differences, but increasingly it seems probable that when my dog Duff sees me packing my suitcase and looks downcast while lying on the suitcase lid, he is feeling the sadness of imminent separation.[21] His feelings of fear or joy are also probably similar to my own, given our similarities in brain and behavior. Feelings of playfulness are likely much the same.

I use the hedge "in some form or other" because I want to allow that there may well be some differences. For example, when my dog Duff feels hungry, his experience may be somewhat different from what I experience. For example, when I feel hungry, I might envisage tomato soup, while he probably envisages raw meat. But then you and I might differ in that regard as well, and it is no big deal so far as having consciousness is concerned. Still, while I might be aware that next week our visitors from the Yukon will arrive, Duff is unlikely to be aware of complex thoughts such as that. On the other hand, his smell world is incomparably richer than mine, so that when we go on a walk, he is very excited by things to which I am smell-blind (well, more correctly, anosmic).

The fact that mammals share a fundamental capacity for sleep and that the brainstem mechanisms regulating sleep are essentially the same across all mammals does strongly suggest that during the waking state there are also fundamental similarities in our experiences—such as being aware of body position, of pains, of feeling hungry or thirsty, of feeling tired or cold.[22] Nevertheless, as we saw in Chapter 8, some philosophers and psychologists have been quite certain that only humans are truly conscious, and only because they have the capacity for language. What are their reasons?

One view is that consciousness is a matter of what one sees or feels or hears, and only an animal that has language can tell us whether they do or do not see or feel or hear something. You can tell me you feel hungry. With my dog, I just have to guess from his behavior. One problem with this is that when you tell me you feel hungry, your speech is, well, just behavior. It is not my direct access to your experience. I do not feel your hunger; I merely hear you say, "I feel hungry." So in that regard, I depend on behavior, whether it is your behavior or that of my dog. Yes, the reply goes, but I can *tell* you I am hungry only if I am conscious. Suppose that is true. What follows? Only that consciousness is necessary for speech; it does not follow that speech is necessary for consciousness.

Another argument is that consciousness is essentially inner speech, and unless you have language, you cannot have inner speech.[23] The main problem here is lack of supporting evidence from brains and behavior. A further problem is that consciousness—mine, anyhow—involves so much more than speech. Indeed, we may experience much for which we have no precise linguistic characterization at all, such as the difference between the smell of cinnamon and the smell of cloves or the difference between feeling energetic and feeling excited, or what an orgasm is like. If we consider that language is essentially a tool for communication, then it is no surprise that consciousness does not require language.[24] As an additional, if tangential, point, many animals communicate quite a lot without something that looks much like a human language. Think of communication between wolves, dolphins, ravens, parrots, and marmosets, for starters.

As we have seen, the leading hypothesis concerning what needs to happen in order to be consciously aware of something specific, such as seeing a bear fishing, is that the signals from the external stimuli need to reach areas toward the front of the brain. There needs to be *global ignition*. Does global ignition always activate language areas? Examining the brain scans gives us no reason to think so. If the signal is a visual one, say of a bear fishing, or an auditory one, such as a whale singing, or if it is a product of the imagination, say a visual image of a massive beanstalk, then the brain scans fail to implicate those areas believed to be important for language. This suggests that language is not an essential part of the awareness of these events. Probably I can be aware of them without talking, and even without talking to myself.

Here is additional new evidence against the hypothesis that language is necessary for consciousness. A patient with a prefrontal tumor was examined before surgery to determine whether the language areas would be impaired if the surgery were carried out. The test consisted of using navigated transcranial magnetic stimulation (nTMS), a technique for safely and transiently disrupting neuronal activity. If speech is arrested by such a technique, then it is presumed that the area stimulated is crucial for speech. The surgeon would then carefully avoid that area.

When speech was arrested using nTMS, did the subject lose consciousness? No.[25] Moreover, this finding is consistent with the earlier large body of data from the pioneering research of neurosurgeon George Ojemann, who tested for language areas prior to surgery using direct electrical stimulation.[26] Patients who showed language disruption did not lose consciousness. Nevertheless, it could be argued that the nTMS case is not decisive. The patient is left-handed, and it remains possible that there is some residual language function in the right hemisphere that was unaffected by the nTMS. Additional experiments using nTMS will soon clarify this matter.

WORKING MEMORY

YOU ARE TOLD the address of a building, and you need to remember it as you drive along an unfamiliar part of town. Knowing that you are apt to forget it, you may rehearse in your mind a few times. Still, after a few minutes, you may discover you have forgotten it, or at least you cannot be sure you can recall it correctly. You knew it for perhaps a minute, then it disappeared from memory. This temporary holding of information is called working memory.

How is it that we can vividly remember something for a few minutes, but only for a few minutes, if we are conscious? Earlier studies recording the activity of single neurons in the prefrontal cortex (specifically, the dorsolateral PFC) indicate that working memory (what you can recall for a minute or two) for a specific event depends on sustained activity in the prefrontal cortex. Activity in specific neurons holds the information during the "wait" interval between learning the information and using it in action. Probably, though this is speculation, unless the central thalamus neurons "juice up" the dorsolateral PFC, those neurons do not code and hold signals in working memory. They are "off duty."

These traits of working memory may help explain Scotty's complete lack of long-term memory for events during his sleepwalking activities. Probably because his prefrontal areas are not activated by the central thalamus, even though his sensory cortices probably are, working memory is not functioning. Moreover, because long-term storage of the particulars of an event (for example, your first driver's license test) also seems to require consciousness of the event when it happened, Scotty has no longer-term memory either.

Optimistically, therefore, you can get a feel for how the conscious-of-anything account might fit well with the conscious-of-*specifically-this* account and how evaluative signals concerning needs and goals must play a role. The optimism inspires concrete experiments, and what is to be hoped is that new ideas for connecting the two accounts will take us further along.

What is the role of attention? What *is* attention? Attention is presumably part of your brain's evaluative business, needed because conscious awareness has a limited capacity. Scientists characterize two kinds of attention. One involves a stimulus that "grabs" your attention—a stimulus such as a loud sound, the sudden smell of smoke, or a bright flash of light. That is the bottom-up variety, and it is clearly related to survival, reflecting when priority must be given to a sudden danger over a current aim. The other is top-down (voluntary), and reflects goals, both short term and long term, as well as preferences. The brain regions involved in this kind of attention are more toward the front of the brain.

When you are engaged in a demanding task, such as removing a splinter from your child's foot, you stay focused on the task, mindful of both the overarching goal of removing the splinter with needle and tweezers and of the sequence of tasks needed to accomplish the goal (first clean the spot with rubbing alcohol, and so forth). You purposefully ignore distractions that would interfere with accomplishing your goal, unless the distraction is really important—fire alarm, for example.

The contents of your conscious state shifts as you pay attention to a conversation and then to the clock and then to the baby in the crib and back to the conversation. As suggested, these shifts in what we experience may rest on fast shifts of linkages between widely separated pools of neurons. The linkages might, it has been speculated, consist of coordinated patterns of activity between highly connected hubs, perhaps regulated by brain rhythms.

What directs top-down attention? How does the brain know

what to ignore and what to focus on? This is not well understood, though it may be pertinent to say that there is no little guy in the brain who uses *his* brain to figure out what you should pay attention to. (An unstoppable regress seems imminent.)

Does consciousness just boil down to paying attention? Probably not, although the two are closely linked. Distinguishing attention from consciousness per se is motivated by the finding that during certain tasks, such as reading, a form of nonconscious attention is at work. How do we know that? When you read, your eyes fixate on a chunk (between about 7 and 20 characters) and then jump to the next chunk. To the reader, it may not seem so, but eye trackers show unequivocally that fixating-then-jumping is actually what your eyes do. Reading is really hopping along the lines, chunk by chunk, however smooth your clever brain makes it seem.

The data indicate that just before the eyes make their next jump (saccade), your nonconscious attention scans the next chunk, selects a meaty word for your center of gaze to light on (a word like *murder*, not a word like *indeed*), and guides the eye movement accordingly. This is attention performing in an anticipatory capacity and almost certainly is a factor in many kinds of complex and skilled movements, such as hunting, cooking, and parenting, when attention shifts to relevant components of the ongoing scene. At most this shows that we can sort of pay attention without awareness. What has not been shown experimentally is that we can be aware without paying attention. Attention, as psychologist Michael Cohen and his colleagues point out, seems to be *necessary*, but not *sufficient*, for consciousness.[27]

MANY QUESTIONS regarding the nature of attention, consciousness, and decision making do not yet have anything close to full answers, but each year brings new details about brain functioning that help to fill in the gaps in the story. For example, it is now well understood that brain wiring is loopy, and loopy architecture means that stored information can be brought to bear on incoming signals, allowing the brain to make fast and complex interpretations. For example, you can immediately recognize something as a car or a bicycle, even when the object is seen at an odd angle, is partially occluded, or is depicted in a cartoon drawing.

You might fret that the *shifting linkages among rich club neurons* or global ignition ideas are pretty sketchy, not providing much in the way of mechanisms. You would be right. The mechanisms whereby coordinated activity of transiently linked neurons are presumed to yield conscious experience are not known. Our ignorance notwithstanding, what is exciting is that neuroscientists now have something that, despite its imprecision, can guide exploration. We can move forward.

If you take a long-range view of the topic of consciousness, it is impressive to realize how far the science has come relative to where it was 50 years ago and how much research at *every* level of brain organization has contributed to the picture now emerging. The study of the timing properties of single-cell responses, of the effects of loops in networks, the nature of large-scale connectivity across the brain—all this and much more go into the theoretical hopper as scientists figure out how the brain makes consciousness. Will a single experimental result be the thread we can pull to unravel the whole sweater? Will there be a Watson and Crick of consciousness? Personally, I doubt it. Instead, I timorously predict, there will be a million little important results—converging, coevolving, and generating ever more new research that will in time bring the bacon home.[28] My caution kicks in when I encounter either one of two sorts of dramatic theories: those that claim to have found *the* secret of consciousness, and those that claim that the brain mechanisms for consciousness can never be found.

Epilogue

Balancing Act

NEUROSCIENCE HAS made remarkable progress in understanding how nervous systems work. The development of new tools and techniques, along with experimental ingenuity, has enabled this progress. How wonderful it is to be able to access journals and blogs and find out what is happening at the cutting edge. Quality varies, however. How do we sort the wheat from the chaff? How do we know whether to believe a dashing economist who refers to himself as Dr. Love and tells us that we need eight hugs a day to be happy?[1]

Unequivocally, we need the media to report scientific discoveries in a way that is both accurate and understandable. Such a feat takes a highly knowledgeable journalist who has the writing talent to put the hay down where the goats can get it. Because scientists who do the research are best situated to understand the significance of a result, they have a special responsibility to help the journalists get the story straight.

Occasionally, however, the scientist informing the journalist succumbs to baser impulses, since the scientist may wish to avail himself of the benefits of having his name prominently in print, and his campus public relations department may encourage journalistic high flying. Awkwardly, or in some cases unblushingly, the scientist may find himself (or herself) cranking up the story here and there to get the journalist's attention. Or the journalist may, for his own self-interested desire for a scoop, allow himself to crank up the story to get a sensational headline and more hits on his blog.

Case in point: the website TechNewsDaily.com ran a story entitled "Can Snoops Hack Your Brain?" Here is an excerpt from the central part of the story:

> Researchers looked for what's called a P300 response, a very distinct brain wave pattern that occurs when one relates to or recognizes something. It would occur, for instance, if you were to look at a picture of your mother, or see your Social Security number written out. While this technology doesn't allow someone else to actively go in and search around in our brains, it's definitely a step in that direction. But for this method to yield any valuable information, many conditions need to be exactly right.

The implication from the story is that hacking your brain is not far off. We are led to believe that there will soon be electronic devices that can tell me what you are thinking as I analyze your brain waves. Allegedly, "it's definitely a step in that direction." This is scary stuff. How plausible is the story? Should I start wearing a tinfoil hat to prevent my brain from being hacked? No, the account is not plausible, and I shall explain why.

We briefly encountered the P300 waveform earlier in our discussion of consciousness, but now I want to explain in a little more detail what it is and what brain waves are, the better to assess the claim at issue.

Suppose I fit out a cap with 30 recording electrodes spread evenly around and place this tight cap on your head and run the electrode signals through a device that tracks the electrical signals and prints out the tracking. This is an electroencephalograph (EEG)—like an electrocardiogram, but for the brain rather than the heart (see Figure 9.1). The EEG is clinically very useful in diagnosing epilepsy, as very unusual waveforms appear during an epileptic seizure, and these unusual waveforms are essentially diagnostic of epilepsy. The origin in the brain of the seizure activity can sometimes be determined by looking at the time when a characteristic epileptic waveform first appears on the EEG.

What do these waves represent? The first important piece of information is that they are the aggregate voltage changes in hundreds of millions of neurons. Can they be correlated with specific thoughts so that you can tell from the recordings whether I am thinking of a cat or a cow? No. Not even close. It is like recording the hubbub of a football crowd. You can tell when it gets louder or softer and whether a goal has been scored, but you cannot tell what Ted in row 85 seat 67 is saying or whether he is saying anything at all. For all you know, Ted might be feeling around on the floor for his dropped car keys. The crowd noise is an aggregate of all the voices in the crowd.

So what is a P300? If you use an EEG and clock the exact time you present a stimulus, you can look at the timing of various ups and downs of the EEG pen relative to the stimulus onset. You will not see anything special emerge unless you *average* these ups and down over many trials. When you look at the averaged, cleaned-up data, you may notice that somewhere around 300 milliseconds after the stimulus onset, there is a positive (hence the *P*) wave. Early studies regarding the P300 indicated that it was quite robust when a subject saw something unexpected. Some researchers speculated that it might be correlated with a memory update. Importantly, the P300 was never shown to be specific to a particular stimulus, such as seeing a cat—just an unexpected stimulus. In studying consciousness, Dehaene and

Changeux suggested that a late component of the P300 wave is seen when a stimulus is conscious rather than subliminal or masked in their particular experiments. They do not claim that the P300 indicates awareness of a picture of a cow rather than that of a cat.

So could you hack my brain using the EEG? I mean, could you tell exactly what I am thinking—for example, about baking oatmeal cookies—using that technique? What is the stimulus whose onset we track, and over what would you average? This makes no sense. We have no reason whatever to think that you can read my mind with an EEG.

Nevertheless, you might reply, we could in principle get a recording device from the FBI and find out what football fan Ted is saying, if anything. Can we not do that with neurons? Here is where the analogy breaks down. Ted's saying "I am betting on the Colts" is the property of a single person. Assume the special recording device will pick that up. My thinking about Osoyoos Lake is not the property of a single neuron. My thinking of Osoyoos Lake will involve being conscious, so there is the neuronal activity supporting being conscious of anything, not to mention all that involved in paying attention to the particular thought. It will also involve drawing on memory, so the neurons that make the memory available to consciousness, wherever they are, will be active. There will be neurons involving visual and somatosensory images that will be active. My thinking of Osoyoos Lake is a distributed effect of hundreds of thousands of neurons, many of which may be involved when I think about Horsefly Lake. So a single electrode, supposing you could put it directly in my brain rather than just recording from the scalp, will not be useful. Oh yes, and how would you know exactly where to put it? You don't. And I would not let you, anyhow.

Here is a further point that makes hacking the brain far-fetched—why we are not even close to being able to have a device that will succeed at that. Even if you and I are twins and we both think about Osoyoos Lake, not exactly the same pattern of activity or even the same neurons will be associated. That is because we acquired our knowledge somewhat differently and because our brains at the microlevel are organized somewhat differently. We have slightly different biochemistry, and there may be somewhat different epigenetic effects that alter the course of how the brain gets wired up. You had a concussion at 12, I had one at 14. Additionally, in my own brain, it is improbable that exactly the same neurons are doing exactly the same thing every time I think of Osoyoos Lake. My brain changes over time, and maybe some of the neurons involved last time have died, maybe others have been recruited, maybe the whole dynamic pattern got changed.

The point of this section is that enthusiasm for results needs to be balanced by questioning. We need to press: How does that work? What exactly is the evidence? Even so, caution and questioning need to be followed up with some authoritative evaluation of the claims. Where can we go for a balanced assessment of some of the scary or controversial claims? Where, short of becoming a full-fledged neuroscientist, can we find out what is myth and what is fact?

Here is one suggestion. The Society for Neuroscience has launched a web page called *Brainfacts.org*. It is a trustworthy site with a full-time staff. It is supported by both the Kavli Foundation and the Gatsby Foundation. Nick Spitzer, a renowned neuroscientist, is editor in chief, and he has assembled an impressive list of editors known for their accomplishments as well as their integrity. One section, called "Ask an Expert," allows you to ask whether what you have been told is rubbish, a gross exaggeration, a mild puffing up of the hard facts, or generally accurate, if surprising. Bear in mind, too, that even within a field such as neuroscience or linguistics, there are often educated disagreements about what exactly a result means. That is because so much, including so much that involves basic principles, is still not nailed tightly down. Neuroscience is a young science. So even the experts on Brainfacts.org may well qualify their answer to your question by acknowledging that there is much that is not known.[2]

The 1998 report by Andrew Wakefield in the journal *Lancet* claiming that autism is caused by the measles-mumps-rubella vaccine was fraudulent and was later retracted by *Lancet*. Wakefield was found guilty of professional misconduct, and his medical license was revoked. There is not one shred of evidence to link autism to the MMR vaccine, as multiple large studies make evident. Yet many parents believed Wakefield's explanation, and many still believe it. Why? Because it offers a simple answer to what is in fact an extremely complicated and utterly heartbreaking medical condition. The desire to have an answer is so strong that it outweighs the countervailing evidence. Conspiracy theories are drummed up to dismiss the evidence challenging the link as well as the facts regarding Wakefield's fraud. Even worse, many parents have chosen not to vaccinate their children, with the unsurprising but tragic result that there have been outbreaks of measles, mumps, and rubella that have caused permanent injuries and in some cases death. This was not a fraud that hurt nobody. Much the same can be said for those who deny that the human immunodeficiency virus (HIV) is causally implicated in AIDS.[3] This is rubbish that caused a lot of damage.

Balance. As Aristotle pointed out, it is all about balance. You do not want to be so skeptical that you learn little and fail to take advantage of scientific progress. You do not want to be so smug that you think you have nothing to learn from science. Only the fervently ignorant can believe that a woman cannot get pregnant if raped,[4] for example, or that oil of oregano can cure whooping cough (a current and dangerous fad). At the same time, you do not want to be so gullible that you assume that any science blog is as good as any other. Harder still, you do not want to selectively believe the results you happen to favor. Cherry picking, as such selectivity is called, is a recipe for self-deception.

Even the leading journals sometimes get it completely wrong. They occasionally publish an astonishing result that turns out not to be replicable and suffers massaged, but camouflaged, statistics.[5] They occasionally reject a profoundly important article that goes against the grain of orthodoxy. No one gets it right every time. But there is no better system than peer review, with the constant reminder that science cannot thrive unless we are hard-assed when we review a result.

If we start getting soft-soapy and mushy-headed, if we give our pals a special break or kowtow to the mighty, the integrity of science erodes. We are obliged to criticize each other, fairly and forthrightly. We are obliged to take criticism gracefully and thoughtfully. We have to take a close look at the methods as well as the results. We must tell the truth. This sometimes entails being uncompromising with beloved scientific colleagues, especially when they have a story to peddle that people passionately want to be true. The hardest part is dispassionately probing a result we fondly hope is true.[6]

GETTING USED TO MY BRAIN

I ENTERED an elevator, joining an anthropologist who had got on earlier. My very presence brought her to fury, and she hissed, "You reductionist! How can you think there is nothing but atoms?" Who, me? I was flabbergasted. It was as though the word *reductionist* was in the same league as *gestapo* or *arsonist* or *hired assassin*. As with the philosopher's lament "I hate the brain," this outburst motivated some inquiry.

Reductionism is often equated with *go-away-ism*—with claiming that some high-level phenomenon does not really exist.

But wait. When we learn that fire really is rapid oxidation—that is the real underlying nature of fire—we do not conclude that fire does not exist. Rather, we understand a macrolevel thing in terms of microlevel parts and their organization. That is a reduction. If we understand that epilepsy is owed to the sudden synchronous firing of a group of neurons, which in turn triggers similar synchronous firing in other areas of the cortex, this is an *explanation* of a phenomenon; it is not denial of the *existence* of a phenomenon. It is a reduction. True enough, this brain-based explanation of epilepsy does supplant the earlier explanations in terms of supernatural sources. But its success is owed to its having vastly greater evidential support.

Some things people have thought actually exist probably do not—like mermaids, vaginal teeth, demonic possession, and animal spirits. Finding out the nature of things sometimes results in discoveries about what does exist or what does not. Some things that humans did not suspect exist do in fact exist. The existence of so many extinct species, including extinct hominins, such as *Homo neanderthalis* and *Homo erectus*, is surely even more amazing than finding out that there are no leprechauns. The existence of radioactivity, unsuspected until its discovery in the late nineteenth century, was cause for astonishment.

If, as seems increasingly likely, dreaming, learning, remembering, and being consciously aware are activities of the physical brain, it does not follow that they are not real. Rather, the point is that their reality depends on a neural reality. If reductionism is essentially about *explanation*, the lament and the lashing out are missing the point.[7] Nervous systems have many levels of organization, from molecules to the whole brain, and research on all levels contributes to our wider and deeper understanding.[8]

Scientism is the label sometimes slapped on those of us who look to evidence when we seek an explanation. This label, too, is meant to be insulting and is apt to be followed up with the accusation that we stupidly assume that science is the only important thing in life, that nothing but science matters.

Woe is me. Of course there is much in life. Of course, science is not the be-all and end-all. Early this morning I paddled a kayak out on the waters around Bowen Island, all calm and quiet and magnificent. A mother seal and her baby were hunting fish for breakfast on the far side of Hutt Island. Our golden retrievers were waiting for me when I beached the kayak later. Last week, unaccountably, a raven flew in the back door and perched on the rocking chair. She was utterly statuesque and stunningly huge. All beauty. I talked to her briefly, and then she found her way out again. Balance, dear Aristotle, balance.

Science is an extension of common sense. It is common sense gone systematic. Einstein put it well: "One thing I have learned in a long life: that all our science, measured against reality, is primitive and childlike—and yet it is the most precious thing we have." Slamming that comment as scientism would be nothing short of silly.

At the risk of repeating myself, I do emphasize that there is much that remains to be discovered about the neural reality of our mental lives. So many deep puzzles remain. But to rail against reality seems to me unproductive. By contrast, what is productive is finding ways to integrate what we do know, finding what connects and what does not, what might need rethinking and what needs a brand-new approach.

Concern about death as a finality naturally motivates resistance to the recognition that our mental life is a product of our brain life. The death of the brain, the facts suggest, entails the death of the mind. If we like life, that takes a bit of getting used to. Not infrequently, imagination may divert us by creating ideas about some other universe where souls go after death. Souls? What could such a soul be, since memory, skill, knowledge, temperament, and feeling all seem to depend on activities of neurons in the brain? Well, somehow. And where could that place be, given what we know about the universe? Well, somewhere. No one knows for certain it cannot be so. Imagination provides some reassurance but conveniently leaves the details hazy.

Such reassurances often fail and may be seen as carrying too high a cognitive price. For one thing, they muddle how we keep our factual accounts, like adding zeros to the left side of the decimal in our bank account. In my experience, children tend to prefer to hear what we truly think and are quick to recognize when they are being offered a soothing tale. When my Sunday school teachers talked about heaven, they really lost all credibility with us farm brats. We figured they were making it up. The story just did not hang together. *Above the moon? There are just stars above the moon and stars are really faraway suns. Who could live there?* Still, human brains compartmentalize rather efficiently, and you may take the uncertainty ("who knows for sure?") as enough to keep you from getting gloomy as you think about death and dying.

It worked out a little better when our Sunday school teachers said that *we* were what lived on when our grandparents and parents died and that we needed to do what we could to make them proud. We needed to build on their achievements or go in new directions that they would find honorable and useful. That made sense to us. Each of us has to come to terms in our own way with the existential facts of life.

I have told you a little about how I see things, but that is only how *I* see things. Longing for heaven and preparing to enter heaven seem much less pressing to me than making a difference here and now. I am more grateful to George Washington and Thomas Jefferson than I am to the monks who spend their time praying for their own souls in hopes of going to heaven. I am more grateful to those who invented safe and effective contraception than I am to those who merely warned that my soul was doomed if I used it. I am more grateful to those working long hours to understand the causes of Alzheimer's disease than I am to those who merely extol the beauty of ignorance.

I had the great good fortune of meeting Jonas Salk shortly after we moved to San Diego in 1984. In 1951, our village had been ravaged by polio. The butcher survived by living in an iron lung for six months, not knowing whether he would ever get out. Three or four children were crippled in various ways and in varying degrees. Some had milder cases with brief durations. We were, one and all, terrified. Fifty years earlier, my father had suffered polio at age 4. He survived the illness, but was left with a withered leg and walked with a limp. Not an easy handicap on a working farm. All this just spilled out of me when I came to know Jonas well and felt comfortable enough to tell him how overwhelmingly much it meant to us when his vaccine became available in 1955.

Bertrand Russell, philosopher and mathematician, has the last word:

Even if the open windows of science at first make us shiver after the cozy indoor warmth of traditional humanizing myths, in the end the fresh air brings vigor, and the great spaces have a splendor of their own.[9]

Rock on, Bertie.

NOTES

Chapter 1: Me, Myself, and My Brain

1 Some of these worries find voice in Christof Koch, *Consciousness: Confessions of a Romantic Reductionist* (Cambridge, MA: MIT Press, 2012). See also the review of this book by Stanislas Dehaene, "The Eternal Silence of Neuronal Spaces," *Science* 336 (2012): 1507–8.

2 Why did Galileo get such grief, but not Copernicus? All too aware of the punishment for heresy, Copernicus did not risk publishing his book and incurring the wrath of the Church while he was alive. Nevertheless, the Church did ban his book, as Copernicus fully expected it would. The banning was futile; the word was out.

3 Galen was certainly no fool. On the contrary, he made many dissections and made significant contributions to anatomy. In cutting nerves in animals, he correctly came to the realization that the nerves from the brain and spinal cord control the muscles. The real problem was that physicians came to treat his work as authoritative and beyond questioning or correction, something Galen himself probably would have discouraged. It was especially unfortunate when so little was understood about how the body works.

4 *De Motu Cordi* (*On the Motion of the Heart and Blood in Animals*), first published 1628 in Frankfurt. Prometheus Books republished a paperback in 1993.

5 Thomas Wright, *William Harvey—A Life in Circulation* (New York: Oxford University Press, 2012).

6 On this topic, see Owen Flanagan, *The Problem of the Soul: Two Visions of the Mind and How to Reconcile Them* (New York: Basic Books, 2002).

7 This is widely attributed to Huxley, but exactly where and when he said it I cannot discover.

8 To be sure, some philosophers have warmly embraced neuroscience, and among those best known, I note Chris Eliasmith, Owen Flanagan, Andy Clark, David Livingston Smith, Clark Glymour, Tim Lane, Jeyoun Park, Brian Keeley, Warren Bickel, Walter Sinott-Armstrong, Carl Craver, Leonardo Ferreira Almada, Thomas Metzinger, Dan Dennett, Ned Block, and Rick Grush. See also Massimo Piglucci's philosophy website, http://rationallyspeaking.blogspot.com/, and Nigel Warburton's Philosophy Bites interviews and podcasts on http://www.philosophybites .com/ and Julian Baggini's website http://www.microphilosophy.net/.

9 O. Flanagan and D. Barack, "Neuroexistentialism," *EurAmerica: A Journal of European and American Studies* (2010): 573–90.

10 The best-known textbook, affectionately referred to as the "*NeuroBible*," is *Principles of Neural Science*, now in its fifth edition, by Eric Kandel et al. (New York: McGraw-Hill Medical, 2012). An outstanding introduction is *Fundamental Neuroscience* (Elsevier, 2003) by Larry Squire et al., now in its fourth edition, but with a new edition in press. Another favorite is *Neuroscience, Fifth Edition*, by Dale Purves et al. (Sunderland, MA: Sinauer, 2011). A very simple but useful introduction is by Sam Wang and Sandra Aamodt, *Welcome to Your Brain* (New York: Bloomsbury, 2008).

11 See Matt Ridley, *The Rational Optimist: How Prosperity Evolves* (New York: Harper, 2010).

12 This is a point first impressed upon me by Brian Cantwell Smith, who at that time worked at Xerox PARC.

Chapter 2: Soul Searching

1 This is well explained by Rodolfo Llinas in *I of the Vortex* (Cambridge, MA: MIT Press, 2001).

2 Daniel Dennett also makes this point in *Consciousness Explained* (Boston: Little, Brown, 1992).

3 For a much fuller account of this hypothesis, see Paul Churchland, *Plato's Camera* (Cambridge, MA: MIT Press, 2012).

4 M. J. Hartmann, "A Night in the Life of a Rat: Vibrissal Mechanics and Tactile Exploration," *Annals of the New York Academy of Sciences* 1225

(2011): 110–18. For a delightful discussion of the capacities of rats to do well in the world, see Kelly Lambert, *The Lab Rat Chronicles* (New York: Perigee, 2011).

5 It turns out that even blindfolded subjects with a 30-centimeter-long "hair" attached to their index finger can quickly learn to perform spatial tasks, but subjects did not report on what the experience was like. See A. Saig, G. Gordon, E. Assa, A. Arieli, and E. Ahissar, "Motor-Sensory Confluence in Tactile Perception," *Journal of Neuroscience* 32, no. 40 (2012): 14022–32. doi: 10.1523/JNEUROSCI.2432-12.2012.

6 See L. Krubitzer, K. L. Campi, and D. F. Cooke, "All Rodents Are Not the Same: A Modern Synthesis of Cortical Organization," *Brain, Behavior and Evolution* 78, no. 1 (2011): 51–93.

7 For this and many other wonderful accounts of crow and raven behavior, see J. Marzluff and T. Angell, *Gifts of the Crow* (New York: Simon and Schuster, 2012).

8 See V. S. Ramachandran and S. Blakeslee, *Phantoms in the Brain* (New York: HarperCollins, 1999).

9 G. Bottini, E. Bisiach, R. Sterzi, and G. Vallar, "Feeling Touches in Someone Else's Hand," *NeuroReport* 13, no. 2 (2002): 249–52.

10 See the excellent discussion of the Bottini case in T. Lane and C. Liang, "Mental Ownership and Higher-Order Thought," *Analysis* 70, no. 3 (2010): 496–501.

11 P. Redgrave, N. Vautrelle, and J. N. Reynolds, "Functional Properties of the Basal Ganglia's Re-entrant Loop Architecture: Selection and Reinforcement," *Neuroscience* 198 (2011): 138–51.

12 I am reluctant to give his name, as this was really one of those post-doc adventures that would now embarrass him. At the time he told me about it, we were both blown away.

13 *The Works of Aristotle, Vol III*, translated by W. D. Ross (Oxford: Clarendon Press, 1931). *De Anima* I 1 402a10-11.

14 *Catholic Encyclopedia, General Resurrection*. Go to http://www .newadvent.org/cathen/12792a.htm.

15 A prominent thinker I am leaving out is Thomas Aquinas (1225–1274), who was greatly influenced by Aristotle but who still had to work into his theology the Christian belief in the resurrection of the body. No mean feat. See the discussions of Aquinas in a special issue of *Philosophical Topics* (vol. 27, no. 1, 1999).

16 For a video of the split-brain patient Joe, go to http://www.youtube .com/watch?v=ZMLzP1VCANo.

17 M. S. Gazzaniga and J. E. LeDoux, *The Integrated Mind* (New York: Plenum Press, 1978).

18 This is true of most neurons, but amazingly enough, the squid has a

very large motor neuron that controls the muscles of its water propulsion system, and that axon can be seen with the naked eye. Owing to its size (about 1 millimeter in diameter), the giant axon of the squid became a useful object for neuroscientists to study in the early stages of neuroscience in the twentieth century. It allowed Hodgkin and Huxley to figure out the mechanism whereby neurons send and receive information.

19 For a terrific website that shows how neurons work, see Gary Matthews, http://www.blackwellpublishing.com/matthews/animate.html. For a more advanced electronic text by Gary Matthews, go to http://books.google.com/books/about/Introduction_to_Neuroscience.html?id=1dRYEQwJcEAC.

20 Edmund Taylor Whittaker, *A History of the Theories of Aether and Electricity: From the Age of Descartes to the Close of the Nineteenth Century* (Forgotten Books: Originally published 1910, reprinted 2012).

21 See Noam Chomsky make this claim: http://www.youtube.com/watch?v=s_FKmNMJDNg&feature=related.

22 Colin McGinn, *The Mysterious Flame: Conscious Minds in a Material World* (New York: Basic Books, 1999).

23 Stuart Firestein, *Ignorance: How It Drives Science* (New York: Oxford University Press, 2012).

24 David Chalmers, *The Conscious Mind* (New York: Oxford University Press, 1997).

25 J. A. Brewer, P. D. Worhunsky, J. R. Gray, Y. Y. Tang, J. Weber, and H. Kober, "Meditation Experience Is Associated with Differences in Default Mode Network Activity and Connectivity," *Proceedings of the National Academy of Sciences USA* 108, no. 50 (2011): 20254–59.

Chapter 3: My Heavens

1 Suze Orman has a television show on money management that airs on Saturday night on CNBC.

2 For the *Brain Death Guidelines for Pediatrics*, see http://pediatrics.aappublications.org/content/early/2011/08/24/peds.2011-1511.

3 There is a continuum of conditions, and it is useful to consult the Glasgow Coma Scale for more precision: http://www.unc.edu/~rowlett/units/scales/glasgow.htm.

4 To see a scan of her brain, go to Google Images and search for Terri Schiavo.

5 For an excellent review article, see Dean Mobbs and Caroline Watt,

"There Is Nothing Paranormal About Near-Death Experiences: How Neuroscience Can Explain Seeing Bright Lights, Meeting the Dead, or Being Convinced You Are One of Them," *Trends in Cognitive Sciences* 15, no. 10 (2011): 447–49.

6 See also Michael N. Marsh, *Out-of-Body and Near-Death Experiences: Brain-State Phenomena or Glimpses of Immortality?* (New York: Oxford University Press, 2010).

7 For a wise and insightful discussion of his drug taking by Oliver Sacks, go to: http://www.newyorker.com/online/blogs/culture/2012/08/video-oliver-sacks-discusses-the-hallucinogenic-mind.html.

8 Pim van Lommel, Ruud van Wees, Vincent Meyers, and Ingrid Elfferich, "Near-Death Experience in Survivors of Cardiac Arrest: A Prospective Study in the Netherlands," *Lancet* 358 (2001): 2041.

9 Oliver Sacks, *Hallucinations* (New York: Knopf, 2012).

10 J. Allan Hobson, *Dreaming: An Introduction to the Science of Sleep* (New York: Oxford University Press, 2003).

11 M. Paciaroni, "Blaise Pascal and His Visual Experiences," *Recenti Progressi in Medicinia* 102, no. 12 (December 2011): 494–96. doi: 10.1701/998.10863.

12 http://ehealthforum.com/health/topic29786.html; http://psychcentral.com/lib/2010/sleep-deprived-nation/.

13 "Brain bugs" is the name that neuroscientist Dean Buonamano gives to events such as hallucinations. (See his book, *Brain Bugs*, 2011.)

14 Gantry is a fictional character in the novel of the same name by Sinclair Lewis.

15 This is generally attributed to Russell, but I cannot track down the source. He wrote voluminously, and lectured frequently.

16 This claim was made by Senator Todd Akin of Missouri and has been followed up by television actor Kirk Cameron, who is trying to set up a clinic in San Francisco for breastfeeding gay men. One difficulty seems to be that few lactating women are willing to provide the service. http://dailycurrant.com/2012/08/29/kirk-cameron-ready-start-curing-gays-breastmilk/.

17 D. Pauly, J. Adler, E. Bennett, V. Christensen, P. Tyedmers, and R. Watson, "The Future for Fisheries," *Science* 21 (2003): 1359–61.

18 T. Lane and O. Flanagan, "Neuroexistentialism, Eudaimonics and Positive Illusions," in Byron Kaldis, ed., *Mind and Society: Cognitive Science Meets the Philosophy of the Social Sciences*, Synthese Library Series (New York: Springer, 2012).

19 See S. Salerno, "Positively Misguided: The Myths and Mistakes of the Positive Thinking Movement," *eSkeptic* (April 2009); http://www

.skeptic.com/eskeptic/09-04-15/#feature. See the response by M. R. Waldman and A. Newburg, e*Skeptic* (May 27, 2009), and Salerno's reply: http://www.skeptic.com/eskeptic/09-05-27/.

Chapter 4: The Brains Behind Morality

1 Based on Joseph Boyden, *The Three Day Road* (New York: Penguin, 2005).

2 Lest this story seem completely set apart from other human conventions, recall this Bible story of David's difficulties (2 Samuel 12:11–14 NAB):

Thus says the Lord: "I will bring evil upon you out of your own house. I will take your wives while you live to see it, and will give them to your neighbor. He shall lie with your wives in broad daylight. You have done this deed in secret, but I will bring it about in the presence of all Israel, and with the sun looking down."

Then David said to Nathan, "I have sinned against the Lord." Nathan answered David: "The Lord on his part has forgiven your sin: you shall not die. But since you have utterly spurned the Lord by this deed, the child born to you must surely die." [The child dies seven days later.]

3 See Chapter 8, "Hot Blood," in Nick Lane's brilliant book *Life Ascending: The Ten Great Inventions of Evolution* (New York: W. W. Norton, 2009), 211.

4 S. R. Quartz and T. J. Sejnowski, "The Constructivist Brain," *Trends in Cognitive Sciences* 3 (1999): 48–57; S. R. Quartz and T. J. Sejnowski, *Liars, Lovers and Heroes* (New York: Morrow, 2003).

5 T. W. Robbins and A. F. T. Arnsten, "The Neuropsychopharmacology of Fronto-Executive Function: Monoaminergic Modulation," *Annual Review of Neuroscience* 32, no. 1 (2009): 267–87.

6 S. J. Karlen and L. Krubitzer, "The Evolution of the Neocortex in Mammals: Intrinsic and Extrinsic Contributions to the Cortical Phenotype," *Novartis Foundation Symposium* 270 (2006): 146–59; discussion 159–69.

7 See L. Krubitzer, "The Magnificent Compromise: Cortical Field Evolution in Mammals," *Neuron* 2 (2007): 201–8; also see L. Krubitzer, K. L. Campi, and D. F. Cooke, "All Rodents Are Not the Same: A Modern Synthesis of Cortical Organization," *Brain, Behavior and Evolution* 78, no. 1 (2011): 51–93.

8 L. Krubitzer and J. Kaas, "The Evolution of the Neocortex in Mammals: How Is Phenotypic Diversity Generated?" *Current Opinion in Neurobiology* 15, no. 4 (2005): 444–53.

9 Y. Cheng, C. Chen, C. P. Lin,, K. H. Chou, and J. Decety, "Love Hurts: An fMRI Study," *NeuroImage* 51 (2010): 923–29.

10 Stephen W. Porges and C. Sue Carter, "Neurobiology and Evolution: Mechanisms, Mediators, and Adaptive Consequences of Caregiving," in *Moving Beyond Self-Interest: Perspectives from Evolutionary Biology, Neuroscience, and the Social Sciences,* ed. S. L. Brown, R. M. Brown, and L. A. Penner (New York: Oxford University Press, 2011.); see also E. B. Keverne, "Genomic Imprinting and the Evolution of Sex Differences in Mammalian Reproductive Strategies," *Advances in Genetics* 59 (2007): 217–43.

11 C. S. Carter, A. J. Grippo, H. Pournajafi-Nazarloo, M. Ruscio, and S. W. Porges, "Oxytocin, Vasopressin, and Sociality," in *Progress in Brain Research* 170, ed. Inga D. Neumann and Rainer Landgraf (New York: Elsevier, 2008), 331–36; L. Young and B. Alexander, *The Chemistry Between Us: Love, Sex and the Science of Attraction* (New York: Current Hardcover, 2012).

12 K. D. Broad, J. P. Curley, and E. B. Keverne, "Mother–Infant Bonding and the Evolution of Mammalian Social Relationships," *Philosophical Transactions of the Royal Society B: Biological Sciences* 361, no. 1476 (2006): 2199–214.

13 F. L. Martel, C. M. Nevison, F. D. Rayment, M. J. Simpson, and E. B. Keverne, "Opioid Receptor Reduces Affect and Social Grooming in Rhesus Monkeys," *Psychoneuroendocrinology* 18 (1993): 307–21; E. B. Keverne and J. P. Curley, "Vasopressin, Oxytocin and Social Behaviour," *Current Opinion in Neurobiology* 14 (2004): 777–83.

14 The changes are complex, and my brief summary is simplified. See the review by A. Veenema, "Toward Understanding How Early-Life Social Experiences Alter Oxytocin- and Vasopressin-Regulated Social Behaviors," *Hormones and Behavior* 61, no. 3 (2012): 304–12.

15 Stephen W. Porges and C. Sue Carter, "Neurobiology and Evolution: Mechanisms, Mediators, and Adaptive Consequences of Caregiving," in *Moving Beyond Self-Interest: Perspectives from Evolutionary Biology, Neuroscience, and the Social Sciences,* ed. S. L. Brown, R. M. Brown, and L. A. Penner (New York: Oxford University Press, 2011.); see also E. B. Keverne, "Genomic Imprinting and the Evolution of Sex Differences in Mammalian Reproductive Strategies," *Advances in Genetics* 59 (2007): 217–43.

16 Jaak Panksepp, "Feeling the Pain of Social Loss," *Science* 302, no. 5643 (2003): 237–39. Note, however, that alligators will defend the nest if the young squeal. Synapsids, or reptile-like mammals, are believed to have branched off from sauropsids (reptiles) about 315 million years

ago. Little is known about the evolution of mammals because mammals are the only surviving synapsids, all other intermediate species having become extinct.

17 P. S. Churchland and P. Winkielman, "Modulating Social Behavior with Oxytocin: How Does It Work? What Does It Do?" *Hormones and Behavior* 61 (2012): 392–99. For a very recent paper that draws attention to the different ways that oxytocin can act in the brains of different species, see James L. Goodson. "Deconstructing Sociality, Social Evolution, and Relevant Nonapeptide Functions." *Psychoneuroendocrinology.* (2013) online www.elsevier.com/locate/psychoneuen. The implication Goodson draws is that the prairie vole, fascinating though it is, may, but may *not*, be a model for human social behavior. We need to know much more about the human. See also: http://neurocritic.blogspot.com/2012/07/paul-zak-oxy tocin-skeptic.html.

18 C. L. Apicella, D. Cesarini, M. Johanneson, C. T. Dawes, P. Lich-tenstein, B. Wallace, J. Beauchamp, and L. Westberg, "No Association Between Oxytocin Receptor (OXTR) Gene Polymorphisms and Experi-mentally Elicited Social Preferences," *PlosOne* 5, no. 6 (2010): e11153.

19 Jay Schulkin, *Adaptation and Well-Being: Social Allostasis* (Cam-bridge, UK: Cambridge University Press, 2011).

20 George P. Murdock and Suzanne F. Wilson, "Settlement Patterns and Community Organization: Cross-Cultural Codes 3," *Ethnology* 11 (1972): 254–95.

21 George P. Murdock and Suzanne F. Wilson, "Settlement Patterns and Community Organization: Cross-Cultural Codes 3," *Ethnology* 11 (1972): 254–95. Frans Boas (1888/1964, p. 171) writes that among the Eskimo he observed, monogamy was much more frequent than polygyny. For genetic evidence of prehistoric polygyny, see I. Dupanloup, L. Pereira, G. Bertorelle, F. Calafell, M. J. Prata, A. Amorim, and G. Barbujani, "A Recent Shift from Polygyny to Monogamy in Humans Is Suggested by the Analysis of Worldwide Y-Chromosome Diversity," *Journal of Molecular Evolution* 57, no. 1 (2003): 85–97.

22 L. Fortunato and M. Archetti, "Evolution of Monogamous Mar-riage by Maximization of Inclusive Fitness," *Journal of Evolutionary Biology* 23, no. 1 (2010): 149–56.

23 L. Fortunato and M. Archetti, "Evolution of Monogamous Mar-riage by Maximization of Inclusive Fitness," *Journal of Evolutionary Biology* 23, no. 1 (2010): 149–56.

24 K. Izuma, "The Social Neuroscience of Reputation," *Neuroscience Research* 72 (2012): 283–88.

25 J. Decety, "The Neuroevolution of Empathy," *Annals of the New York Academy of Sciences* 1231 (2011): 35–45.

26 See S. Blackburn, *Being Good: A Short Introduction* (New York: Oxford University Press, 2001); P. S. Kitcher, *The Ethical Project* (Cambridge, MA: Harvard University Press, 2012).

27 See also Dale Peterson, *The Moral Lives of Animals* (New York: Bloomsbury Press, 2011).

28 Not his real name.

29 At least as I remember it, and memory can be faulty.

30 For a witty book on the Scots in Canada, see John Kenneth Galbraith, *The Scotch* (Boston: Houghton Mifflin, 1985).

31 As Frank Bruni recently pointed out ("Suffer the Children," *New York Times*, Sept. 11, 2012), these sentiments are what allowed convicted pedophile Jerry Sandusky to continue to abuse boys at Penn State and what allowed Father Ratigan of Kansas City to continue to molest children.

32 P. S. Kitcher, *The Ethical Project* (Cambridge, MA: Harvard University Press, 2012).

33 E. A. Hoebel, *The Law of Primitive Man: A Study in Comparative Legal Dynamics* (Cambridge, MA: Harvard University Press, 2006), Chapter 5.

34 In modern times, we would hope that government policy would be highly responsive to whether the policy achieves what it aims for. In fact, Ben Goldacre argues, this is seldom done. See his essay: http://www.badscience.net/2012/06/heres-a-cabinet-office-paper-i-co-authored-about-randomised-trials-of-government-policies/#more-2524.

35 See the discussion by Christine Korsgaard, "Morality and the Distinctiveness of Human Action," in *Primates and Philosophers: How Morality Evolved*, ed. Frans de Waal, Stephen Macedo, and Josiah Ober (Princeton, NJ: Princeton University Press, 2006), 98–119.

36 See the video of capuchin reaction to unfairness in de Waal's TED talk: http://www.youtube.com/watch?v=-KSryJXDpzo

37 Sarah F. Brosnan, "How Primates (Including Us!) Respond to Inequity," in *Neuroeconomics* (*Advances in Health Economics and Health Services Research*, vol. 20), ed. Daniel Houser and Kevin McCabe (Bingley, UK: Emerald Group, 2008), 99–124.

38 In conversation.

39 See J. Decety, G. J. Norman, G. G. Berntson, and J. T. Cacioppo, "Neurobehavioral Evolutionary Perspective on the Mechanisms Underlying Empathy," *Progress in Neurobiology* 98, no. 1 (2012): 38–48.

40 See Lynn Hunt, *Inventing Human Rights: A History* (New York: W. W. Norton, 2007).

41 See D. L. Everett, *Don't Sleep, There are Snakes: Life and Language in the Amazonian Jungle* (New York: Pantheon, 2009).

42 See again E. A. Hoebel, *The Law of Primitive Man: A Study in*

Comparative Legal Dynamics (Cambridge, MA: Harvard University Press, 2006), Chapter 5.

43 They are also known as belonging to a larger tribe called the Igorot, and inhabit one province in the mountains.

44 See E. A. Hoebel, *The Law of Primitive Man: A Study in Comparative Legal Dynamics* (Cambridge, MA: Harvard University Press, 2006), 104.

45 See E. A. Hoebel, *The Law of Primitive Man: A Study in Comparative Legal Dynamics* (Cambridge, MA: Harvard University Press, 2006), 104.

46 Loyal Rue, *Religion Is Not About God: How Spiritual Traditions Nurture Our Biological Nature and What to Expect When They Fail* (New Brunswick, NJ: Rutgers University Press, 2005).

47 R. N. Bellah, *Religion in Human Evolution: From the Paleolithic to the Axial Age* (Cambridge, MA: Harvard University Press, 2011); Brian Morris, *Religion and Anthropology: A Critical Introduction* (Cambridge, UK: Cambridge University Press, 2006); K. Armstrong, *A History of God: The 4000-Year Quest of Judaism, Christianity and Islam* (New York: Ballantine Books, 1993).

48 Scott Atran, *In Gods We Trust* (New York: Oxford University Press, 2002).

49 M. Killen and J. Smetana, "Moral Judgment and Moral Neurosciences: Intersections, Definitions, and Issues," *Child Development Perspectives* 2, no. 1 (2008): 1–6; Y. Park and M. Killen, "When Is Peer Rejection Justifiable? Children's Understanding Across Two Cultures," *Cognitive Development* 25 (2010): 290–301.

50 B. Schwartz and K. Sharpe, *Practical Wisdom: The Right Way to Do the Right Thing* (New York: Riverhead Books, 2010).

51 This is explicitly drawn from my book *Braintrust: What Neuroscience Tells Us About Morality* (Princeton, NJ: Princeton University Press, 2011), p. 9.

Chapter 5: Aggression and Sex

1 Also acquitted was one Hispanic policeman.

2 See http://www.youtube.com/watch?v=PmsKGhLdZuQ&feature=related.

3 See http://www.youtube.com/watch?v=M1W_sfJant8; also see http://www.youtube.com/watch?v=9loSB5zMaoQ&feature=related.

4 See http://www.youtube.com/watch?v=M1W_sfJant8.

5 Jonathan Gottschall's most recent book is *The Storytelling Animal: How Stories Make Us Human* (Boston: Houghton Mifflin Harcourt, 2012). He is now working on a book on male violence.

6 D. Lin et al., "Functional Identification of an Aggression Locus in the Mouse Hypothalamus," *Nature* 470 (2011): 221–26.

7 Jaak Panksepp, *Affective Neuroscience* (New York: Oxford University Press, 1998).

8 Jaak Panksepp, *Affective Neuroscience* (New York: Oxford University Press, 1998).

9 See the video on National Geographic: http://video.national geographic.com/video/animals/mammals-animals/dogs-wolves-and -foxes/wolves_gray_hunting/.

10 See Read Montague, *Why Choose This Book: How We Make Decisions* (New York: Dutton, 2006).

11 See the female grizzly fend off the wolves who are trying to get her baby: http://video.nationalgeographic.com/video/animals/mammals -animals/dogs-wolves-and-foxes/wolves_gray_hunting/.

12 See http://www.youtube.com/watch?v=5hw4iRcWbV4.

13 As novelist D. A. Serra pointed out in conversation.

14 Craig B. Stanford, *Chimpanzee and Red Colobus: The Ecology of Predator and Prey* (Cambridge, MA: Harvard University Press, 2001).

15 Massimo Pigliucci and Jonathan Kaplan, *Making Sense of Evolution: The Conceptual Foundations of Evolutionary Biology* (Chicago: University of Chicago Press, 2006).

16 Matt Ridely, *The Red Queen: Sex and the Evolution of Human Nature.* (New York: HarperCollins, 1993); Hans Kummer, *Primate Societies: Group Techniques of Ecological Adaptation* (Chicago: Aldine-Atherton, 1971); Shirley Strum and Linda-Marie Fedigan, *Primate Encounters: Models of Science, Gender, and Society* (Chicago: University of Chicago Press, 2000).

17 Richard Dawkins, *The Blind Watchmaker* (New York: W. W. Norton, 1996).

18 See http://wn.com/Greater_Sage-Grouse_Booming_on_Lek.

19 Many books and articles claim to identify the differences between male and female brains. Some are embarrassingly uninformed, some merely reaffirm old myths in neurospeak, and some do not know the difference between a hormone and a hairpin. One book that really stands out from the crowd is by Donald Pfaff, a neuroendocrinologist at Rockefeller University who has researched the interaction of hormones and the brain all his life. He has written a brilliant book, *Man and Woman: An Inside Story* (New York: Oxford University Press, 2011). Luckily, it is also wonderfully readable. I have drawn heavily on Pfaff's research.

20 See Donald Pfaff, *Man and Woman: An Inside Story* (New York: Oxford University Press, 2011).

21 See Donald Pfaff, *Man and Woman: An Inside Story* (New York: Oxford University Press, 2011).

22 Ai-Min Bao and Dick F. Swaab, "Sexual Differentiation of the Human Brain: Relation to Gender Identity, Sexual Orientation and Neuropsychiatric Effects," *Frontiers in Neuroendocrinology* 32 (2011): 214–26. This is a technical review, but the writing is exceptionally clear and the figures are very helpful.

23 Jan Morris, *Conundrum* (London: Faber, 1974).

24 Estrella R. Montoya, David Terburg, Peter A. Bos, and Jack van Honk, "Testosterone, Cortisol and Serotonin as Key Regulators of Social Aggression: A Review and Theoretical Perspective," *Motivation and Emotion* 36, no. 1 (2012): 65–73; see also Cade McCall and Tania Singer, "The Animal and Human Neuroendocrinology of Social Cognition, Motivation and Behavior," *Nature Neuroscience* 15, no. 5 (2012): 681–88.

25 Jack van Honk, Jiska S. Peper, and Dennis J.L.G. Schutter, "Testosterone Reduces Unconscious Fear but Not Consciously Experienced Anxiety: Implications for Disorders of Fear and Anxiety," *Biological Psychiatry* 58 (2005): 218–25.

26 David Terburg, Barak Morgan, and Jack van Honk, "The Testosterone-Cortisol Ratio: A Hormonal Marker for Proneness to Social Aggression," *International Journal of Law and Psychiatry* 32 (2009): 216–23.

27 This and other puzzling claims about oxytocin can be found in Paul Zak, *The Moral Molecule: The Source of Love and Prosperity* (New York: Penguin, 2012).

28 Anne Campbell, "Attachment, Aggression and Affiliation: The Role of Oxytocin in Female Social Behavior," *Biological Psychiatry* 77, no. 1 (2008): 1–10.

29 Jaak Panksepp, *Affective Neuroscience* (New York: Oxford University Press, 1998).

30 A. R. Damasio, *The Feeling of What Happens* (New York: Harcourt Brace & Company, 1999).

31 Michael Eid and Ed Diener, "Norms for Experiencing Emotions in Different Cultures: Inter- and Intranational Differences," *Journal of Personality and Social Psychology* 81, no. 5 (2001): 869–85.

32 Murder of one man by another, usually over a woman, does happen, and while frowned upon, it is tolerated so long as it does not escalate.

33 Franz Boas, *The Central Eskimo* (Washington, DC: Sixth Annual Report of the Bureau of Ethnology, Smithsonian Institution, 1888).

34 N. Chagnon, *Yanomamo* (New York: Harcourt Brace & Company, 1997).

35 See http://www.rosala-viking-centre.com/vikingships.htm.

Chapter 6: Such a Lovely War

1 This title is adapted from the brilliant stage musical *Oh What a Lovely War*, which was based on the radio show (1961) written by Charles Chilton, *The Long, Long Trail*. It is a chilling contrast of the dim-witted arrogance and optimism of the generals and politicians with the ugliness and horror of the First World War. The musical lampoons the cheery propaganda used to mask the horrors.

2 Chris Hedges, *War Is a Force That Gives Us Meaning* (New York: Public Affairs, 2002), 3.

3 Writing is believed to have emerged in Mesopotamia about 3200 BCE and in Mesoamerica about 600 BCE. For pictures and explanations of Inuit tools, see Franz Boas, *The Central Eskimo* (Washington, DC: Sixth Annual Report of the Bureau of Ethnology, Smithsonian Institution, 1888).

4 Matthew White explores the hypothesis that this may be true of the Aztecs, who did not herd and had no ready source of large, undomesticated animals. The hundreds of thousands of human "sacrifices" made each year, he reports, may have served multiple purposes, including communications with the gods, but also ritualistic food preparation. See Matthew White, *The Great Big Book of Horrible Things: The Definitive Chronicle of History's 100 Worst Atrocities* (New York: W. W. Norton, 2012).

5 For an insightful discussion of how institutional norms can change, see Kwame Anthony Appiah, *The Honor Code: How Moral Revolutions Happen* (New York: W. W. Norton, 2010).

6 For a wonderfully engaging but brief look at the history of the worst wars and massacres, in terms of body count, see Matthew White's book *The Great Big Book of Horrible Things: The Definitive Chronicle of History's 100 Worst Atrocities* (New York: W. W. Norton, 2012). Of course, this is a very disturbing history. Yet the facts are presented so as to provoke reflection, not make you avoid the book altogether.

7 Ian Morris, *Why the West Rules—For Now: The Patterns of History and What They Reveal About the Future* (New York: Farrar, Straus and Giroux, 2010).

8 See also David Livingston Smith, *The Most Dangerous Animal* (New York: St. Martin's Press, 2007).

9 It's also been observed with neuropeptide F, but for simplicity, we can leave that aside here. For more detail, see my book *Braintrust: What Neuroscience Tells Us About Morality* (Princeton, NJ: Princeton University Press, 2011).

10 For a readable and scientifically strong introduction, see Jonathan

Flint, Ralph J. Greenspan, and Kenneth S. Kendler, *How Genes Influence Behavior* (New York: Oxford University Press, 2010). For articles making a related point, see N. Risch and K. Merikangas, "The Future of Genetic Studies of Complex Human Diseases, *Science* 273 (1996): 1516–17; H. M. Colhoun, P. M. McKeigue, and G. D. Smith, "Problems of Reporting Genetic Associations with Complex Outcomes," *Lancet* 361 (2003): 865–72; A. T. Hattersley and M. I. McCarthy, "What Makes a Good Genetic Association Study?" *Lancet* 366 (2005): 1315–23.

11 Herman A. Dierick and Ralph J. Greenspan, "Molecular Analysis of Flies Selected for Aggressive Behavior," *Nature Genetics* 38, no. 9 (2006): 1023–31.

12 This does not necessarily mean that all 80 genes are related to the behavioral phenotype in question, since the differences in some genes may be due to their "hitchhiking" along with those that were selected.

13 See online Supplementary Table 1 for the Dierick and Greenspan (2006) article. doi: 10.1038/ng1864. It is a dramatic demonstration of the results, showing the 80 genes whose expressions were different between the aggressive and neutral flies. Incidentally, since testosterone has often been linked to aggression, it is worth noting that fruit flies do not have testosterone but can be highly aggressive nonetheless.

14 Jonathan Haidt, *The Righteous Mind: Why Good People Are Divided by Politics and Religion* (New York: Pantheon, 2012).

15 Dennis L. Murphy, Meredith A. Fox, Kiara R. Timpano, Pablo Moya, Renee Ren-Patterson, Anne M. Andrews, Andrew Holmes, Klaus-Peter Lesch, and Jens R. Wendland, "How the Serotonin Story Is Being Rewritten by New Gene-Based Discoveries Principally Related to *SLC6A4*, the Serotonin Transporter Gene, Which Functions to Influence All Cellular Serotonin Systems," *Neuropharmacology* 55, no. 6 (2008): 932–60.

16 R. J. Greenspan, "E Pluribus Unum, Ex Uno Plura: Quantitative and Single-Gene Perspectives on the Study of Behavior," *Annual Review of Neuroscience* 27 (2004): 79–105.

17 To get a feel for the difficulty, see, for example, V. A. Vasil'ev, "Molecular Psychogenetics of Deviant Aggressive Behavior in Humans," *Genetika* 47, no. 9 (2011): 1157–68.

18 For a helpful and cautionary article concerning genes and outgroup hostility, see K. G. Ratner and J. T. Kubota, "Genetic Contributions to Intergroup Responses: A Cautionary Perspective," *Frontiers in Human Neuroscience* 6 (2012): 223.

19 See, for example, Jonathan Haidt, *The Righteous Mind: Why Good People Are Divided by Politics and Religion* (New York: Pantheon, 2012).

20 Robert Richardson, *Evolutionary Psychology as Maladapted Psychology* (Cambridge, MA: MIT Press, 2007).

21 For a wise book on this matter, see Stuart Firestein, *Ignorance: How It Drives Science* (New York: Oxford University Press, 2012).

22 Jonathan Haidt, *The Righteous Mind: Why Good People Are Divided by Politics and Religion* (New York: Pantheon, 2012).

23 Among contemporary philosophers, Alasdair MacIntyre understood this as well as anyone. For his classic discussion, set in an Aristotelean framework, see *After Virtue: A Study in Moral Theory*, 3rd ed. (Notre Dame, IN: University of Notre Dame Press, 2007). A more recent and thoughtful exploration, drawing on John Dewey's great work, is Philip Kitcher, *The Ethical Project* (Cambridge, MA: Harvard University Press, 2011).

24 Paul Seabright, *In the Company of Strangers: A Natural History of Economic Life* (Princeton, NJ: Princeton University Press, 2010).

25 Peggy starts out conforming to the secretary role but changes over time, partly owing to her disillusionment with that role after having a baby by one of the up-and-coming admen. The contrast between her copywriting job and the servile work of the secretaries makes their conformity to the secretary role even more striking.

26 See Dov Cohen and Richard E. Nisbett, "Field Experiments Examining the Culture of Honor: The Role of Institutions in Perpetuating Norms About Violence," *Personality and Social Psychology Bulletin* 23 (1997): 1188–99. For a very useful textbook, see *Social Psychology*, 2nd ed., ed. T. Gilovich, D. Keltner, and R. E. Nisbett (New York: W. W. Norton, 2010).

27 David Livingston Smith, *Less Than Human: Why We Demean, Enslave and Exterminate Others* (New York: St. Martin's Press, 2011).

28 Philip Zimbardo, *The Lucifer Effect: Understanding How Good People Turn Evil* (New York: Random House, 2007). The Zimbardo experiment is one of the most important ever done in social psychology. This is because it addressed, carefully and systematically, one of the most painful puzzles of human behavior: How it is that good people can do unspeakably evil things?

29 For a movie that particularly dwells on this, see *The Boy in the Blue Pajamas*.

30 Jonathan Glover, *Humanity* (New Haven, CT: Yale University Press, 2001), 404.

31 Jonathan Glover, *Humanity* (New Haven, CT: Yale University Press, 2001), 281.

32 Steven Pinker, *The Better Angels of Our Nature: Why Violence Has Declined* (New York: Viking, 2012). For a less optimistic view, see David

Livingston Smith, *Less Than Human: Why We Demean, Enslave and Exterminate Others* (New York: St. Martin's Press, 2011).

33 Sam Harris, *Free Will* (New York: Free Press, 2012).

Chapter 7: Free Will, Habits, and Self-Control

1 See also R. F. Baumeister and J. Tienery, *Will Power: Rediscovering the Greatest Human Strength* (New York: Penguin, 2011).

2 C. D. Frith and U. Frith, "How We Predict What Other People Are Going to Do," *Brain Research* 1079, no. 1 (2006): 36–46; R. Adophs, "How Do We Know the Minds of Others? Domain Specificity, Simulation and Enactive Social Cognition," *Brain Research* 1079, no. 1 (2006): 25–35; P. S. Churchland, *Braintrust: What Neuroscience Tells Us About Morality* (Princeton, NJ: Princeton University Press, 2011).

3 See W. Mischel, Y. Shoda, and M. I. Rodriguez, "Delay of Gratification in Children," *Science* 244 (1989): 933–38.

4 B. J. Casey, Leah H. Somerville, Ian H. Gotlibb, Ozlem Aydukc, Nicholas T. Franklina, Mary K. Askrend, John Jonidesd, Marc G. Bermand, Nicole L. Wilsone, Theresa Teslovicha, Gary Gloverf, Vivian Zayasg, Walter Mischel, and Yuichi Shodae, "Behavioral and Neural Correlates of Delay of Gratification 40 Years Later," *Proceedings of the National Academy of Sciences USA* 108, no. 36 (2011): 14998–15003.

5 K. Vohs and R. Baumeister, eds., *The Handbook of Self-Regulation: Research, Theory, and Applications*, 2nd ed. (New York: Guilford Press, 2011); A. Diamond, *Developmental Cognitive Neuroscience* (New York: Oxford University Press, 2012).

6 This simplifies the experimental protocol but gets the essentials. For the accurate details, see R. N. Cardinal, T. W. Robbins, and B. J. Everitt, The Effects of d-Amphetamine, Chlordiazepoxide, Alpha-flupenthixol and Behavioural Manipulations on Choice of Signalled and Unsignalled Delayed Reinforcement in Rats, *Psychopharmacology (Berl.)* 152 (2000): 362–75.

7 See this at http://www.youtube.com/watch?v=kdTdp7Ep6AM, from National Geographic.

8 See http://www.youtube.com/watch?v=SlWe7cO7ThQ&NR=1&feature=fvwp.

9 B. Oakley, A. Knafo, G. Madhavan, and D. D. Wilson, eds., *Pathological Altruism* (New York: Oxford University Press, 2012).

10 See N. Swan, N. Tandon, T. A. Pieters, and A. R. Aron, "Intracranial

Electroencephalography Reveals Different Temporal Profiles for Dorsal- and Ventro-Lateral Prefrontal Cortex in Preparing to Stop Action," *Cerebral Cortex* (published online August 2012. doi: 10.1093/cercor/bhs245); G. Tabibnia et al., "Different Forms of Self-Control Share a Neurocognitive Substrate," *Journal of Neuroscience* 31, no. 13 (2011): 4805–10.

11 S. R. Chamberlain, N. A. Fineberg, A. D. Blackwell, T. W. Robbins, and B. J. Sahakian, "Motor Inhibition and Cognitive Flexibility in Obsessive-Compulsive Disorder and Trichotillomania," *American Journal of Psychiatry* 163, no. 7 (2006): 1282–84.

12 P. S. Churchland and C. Suhler, "Me and My Amazing Old-Fangled Reward System," in *Moral Psychology, Volume 4: Free Will and Moral Responsibility*, ed. Walter Sinnott-Armstrong (Cambridge, MA: MIT Press, 2013).

13 S. M. McClure, M. K. York, and P. R. Montague, "The Neural Substrates of Reward Processing in Humans: The Modern Role of fMRI," *The Neuroscientist* 10, no. 3 (2004): 260–68.

14 The philosopher Eddy Nahmias (http://www2.gsu.edu/~phlean/) decided to find out what ordinary people (that is, nonphilosophers) think about free will. Some philosophers were astounded when he showed that ordinary people have an ordinary conception of free will, more in tune with legal concepts than with refined philosophical concepts, such as those of Descartes or Kant. See E. Nahmias, S. Morris, T. Nadelhoffer, and J. Turner, "Surveying Freedom: Folk Intuitions About Free Will and Moral Responsibility," *Philosophical Psychology* 18, no. 5 (2005): 561–84.

15 I am relying on my own conversations with nonphilosophers, as well as on Eddy Nahmias's more systematic work.

16 G. Lakoff and M. Johnson, *Philosophy in the Flesh* (Cambridge, MA: MIT Press, 1999).

17 Named for Gilles de la Tourette, a clinician who described the disorder in the nineteenth century.

18 This was documented on a CBC program, *The Journal*, and also in a *New Yorker* article, "A Surgeon's Life," by Oliver Sacks (March 16, 1992). Sacks used the name Carl Bennett for Doran.

19 J. F. Leckman, M. H. Bloch, M. E. Smith, D. Larabi, and M. Hampson, "Neurobiological Substrates of Tourette's Disorder," *Journal of Child and Adolescent Psychopharmacology* 20, no. 4 (2012): 237–47.

20 See, for example, Sam Harris, *Free Will* (New York: Free Press, 2012).

21 E. Nahmias, S. Morris, T. Nadelhoffer, and J. Turner, "Surveying Freedom: Folk Intuitions About Free Will and Moral Responsibility," *Philosophical Psychology* 18, no. 5 (2005): 561–84.

22 Daniel Dennett, *Freedom Evolves* (New York: Penguin, 2003). See also my discussion in *Brain-Wise: Studies in Neurophilosophy* (Cambridge, MA: MIT Press, 2002).

23 J. W Dalley, B. J. Everitt, and T. W. Robbins, "Impulsivity, Compulsivity, and Top-Down Control," *Neuron* 69, no. 4 (February 2011): 680–94.

24 Paul Seabright, *The Company of Strangers: A Natural History of Economic Life* (Princeton, NJ: Princeton University Press, 2010).

25 Congress adopted the Insanity Defense Reform Act in 1984. This is what it provides:

(a) Affirmative Defense: It is an affirmative defense to a prosecution under any Federal statute that, at the time of the commission of the acts constituting the offense, the defendant, as a result of severe mental disease or defect was unable to appreciate the nature and quality or the wrongfulness of his acts. Mental disease or defect does not otherwise constitute a defense.

(b) Burden of Proof: The defendant has the burden of proving the defense of insanity by clear and convincing evidence.

On the shifting legal definitions of insanity, see Robert D. Miller, "Part 3: Forensic Evaluation and Treatment in the Criminal Justice System," in *Principles and Practices of Forensic Psychiatry*, 2nd ed., ed. Richard Posner (London: Arnold, 2003), 183–245.

26 C. Bonnie, J. C. Jeffries, and P. W. Low, *A Case Study in the Insanity Defense: The Trial of John W. Hinckley, Jr.*, 3rd ed. (New York: Foundation Press, 2008).

27 A. Caspi, J. McClay, T. E. Moffitt, J. Mill, J. Martin, I. W. Craig, A. Taylor, and R. Poulton, "Role of Genotype in the Cycle of Violence in Maltreated Children," *Science* 297, no. 5582 (2002): 851–54; C. Aslund, N. Nordquist, E. Comasco, J. Leppert, L. Oreland, and K. W. Nilsson, "Maltreatment, MAOA, and Delinquency: Sex Differences in Gene–Environment Interaction in a Large Population-Based Cohort of Adolescents, *Behavior Genetics* 41 (2011): 262–72.

28 http://www.nature.com/news/2009/091030/full/news.2009.1050.html.

29 http://www.tncourts.gov/courts/court-criminal-appeals/opinions/2011/10/20/state-tennessee-v-davis-bradley-waldroup-jr.

30 E. Parens, *Genetic Differences and Human Identities: On Why Talking About Behavioral Genetics Is Important and Difficult*, Hastings Center Report Special Supplement 34, no. 1 (2004).

31 Matthew L Baum, "The Monoamine Oxidase A (MAOA) Genetic Predisposition to Impulsive Violence: Is It Relevant to Criminal Trials?" *Neuroethics* 3 (2011): 1–20. doi: 10.1007/s12152-011-9108-6.

32 Geneticist Ralph Greenspan pointed this out to me.

33 M. Schmucker and F. Lösel, "Does Sexual Offender Treatment Work? A Systematic Review of Outcome Evaluations," *Psicothema* 20, no. 1 (2008): 10–19. See also R. Lamade, A. Graybiel, and R. Prentky, "Optimizing Risk Mitigation of Sexual Offenders: A Structural Model," *International Journal of Law and Psychiatry* 34, no. 3 (2011): 217–25.

34 Regarding why someone might elect castration as a treatment, see L. E. Weinberger, S. Sreenivasan, T. Garrick, and H. Osran, "The Impact of Surgical Castration on Sexual Recidivism Risk Among Sexually Violent Predatory Offenders," *American Academy of Psychiatry and the Law* 33, no. 1 (2005): 16–36.

35 Note that surgically intervening in the nervous system to change social behavior raises further difficult and historically delicate questions, including those concerning safeguards against "treating" the merely eccentric or unusual or obstreperous. See E. S. Valenstein, *Great and Desperate Cures: The Rise and Decline of Psychosurgery and Other Radical Treatments for Mental Illness* (New York: Basic Books, 1986). See also Theo van der Meer, "Eugenic and Sexual Folklores and the Castration of Sexual Offenders in the Netherlands (1938–68)," *Studies in History and Philosophy of Science Part C: Studies in the History and Philosophy of Biological and Biomedical Sciences* 39, no. 2 (2008): 195–204.

36 See J. M. Burns and R. H. Swerdlow, "Right Orbitofrontal Tumor with Pedophilia Symptom and Constructional Apraxia Sign," *Archives of Neurology* 60, no. 3 (2003): 437–440. doi: 10.1001/archneur.60.3.437.

37 D. S. Weisberg, F. C. Keil, J. Goodstein, E. Rawson, and J. R. Gray, "The Seductive Allure of Neuroscience Explanations," *Journal of Cognitive Neuroscience* 20, no. 3 (2008): 470–77.

38 Herbert Weinstein strangled his wife and then threw her out the window of their New York City apartment. For a brief discussion of this case and the significance of Mr. Weinstein's cyst in his defense, see Jeffrey Rosen, "The Brain on the Stand," *New York Times*, March 11, 2007.

Chapter 8: Hidden Cognition

1 T. L. Chartrand and J. A. Bargh, "The Chameleon Effect: The Perception-Behavior Link and Social Interaction," *Journal of Personality*

and Social Psychology 76 (1999): 893–910; J. Lakin and T. L. Chartrand, "Using Nonconscious Behavioral Mimicry to Create Affiliation and Rapport," *Psychological Science* 14 (2003): 334–39.

2 See also M. Earls, M. J. O'Brien, and A. Bentley, *I'll Have What She's Having: Mapping Social Behavior* (Cambridge, MA: MIT Press, 2011).

3 David Livingston Smith, "'Some Unimaginable Substratum': A Contemporary Introduction to Freud's Philosophy of Mind," in *Psychoanalytic Knowledge*, ed. Man Cheung Chung and Colin Fletham (London: Palgrave Macmillan, 2003), 54–75.

4 D. Reiss, *The Dolphin in the Mirror: Exploring Dolphin Minds and Saving Dolphin Lives* (Boston: Houghton Mifflin Harcourt, 2011).

5 B. Heinrich, *Mind of the Raven: Investigations and Adventures with Wolf-Birds* (New York: HarperCollins, 2000).

6 See D. C. Dennett, "Animal Consciousness: What Matters and Why," *Social Research* 62, no. 3 (1995): 691–720. http://instruct.westvalley .edu/lafave/dennett_anim_csness.html.

7 D. C. Dennett, "Animal Consciousness: What Matters and Why," *Social Research* 62, no. 3 (1995): 691–720.

8 To see the evolutionary explanation for why consciousness has its roots in feelings and emotions, see Antonio Damasio and Gil B. Carvalho, "The Nature of Feelings: Evolutionary and Neurobiological Origins." *Nature Reviews Neuroscience* 14 (2013) 143–52 doi: 10.1038/nrn3403. See also J. Panksepp, "Affective Consciousness in Animals: Perspectives on Dimensional and Primary Process Emotion Approaches, "*Proceedings of the Royal Society* 277, no. 1696 (2010): 2905–7.

9 Garth Stein, *The Art of Racing in the Rain* (New York: HarperCollins, 2009).

10 Robert H. Wurtz, Wilsaan M. Joiner, and Rebecca A. Berman, "Neuronal Mechanisms for Visual Stability: Progress and Problems," *Philosophical Transactions of the Royal Society B* 366, no. 1564 (February 2011): 492–503. doi: 10.1098/rstb.2010.0186, PMCID: PMC3030829.

11 P. Haggard and V. Chambon, "Sense of Agency," *Current Biology* 22, no. 10 (2012): R390–92.

12 P. Brugger, "From Haunted Brain to Haunted Science: A Cognitive Neuroscience View of Paranormal and Pseudoscientific Thought," in *Hauntings and Poltergeists: Multidisciplinary Perspectives*, ed. J. Houran and R. Lange (Jefferson, NC: MacFarlane, 2001), 195–213.

13 L. Staudenmaier, *Die Magie als Experimentelle Naturwissenschaft* [Magic as an Experimental Natural Science] (Leipzig, Germany: Akademische Verlagsgesellschaft, 1912), 23.

14 That failures involving predictions of the sequelae of one's actions

may also be a factor in schizophrenia also looks probable. See M. Voss, J. Moore, M. Hauser, J. Gallinat, A. Heinz, and P. Haggard, "Altered Awareness of Action in Schizophrenia: A Specific Deficit in Prediction of Action Consequences," *Brain* 133, no. 10 (2010): 3104–12.

15 For a careful, balanced account of facilitated communication, see the PBS program *Frontline*: http://www.youtube.com/watch?v=DqhlvoU ZUwY&feature=related.

16 P. Simpson, E. Kaul, and D. Quinn, "Cotard's Syndrome with Catatonia: A Case Discussion and Presentation," *Psychosomatics* (in press). http://dx.doi.org/10.1016/j.psym.2012.03.004.

17 D. Guardia, L. Conversy, R. Jardri, G. Lafargue, P. Thomas, P. V. Dodin, O. Cottencin, and M. Luya, "Imagining One's Own and Someone Else's Body Actions: Dissociation in Anorexia Nervosa," *PLOS ONE* 7, no. 8 (2012): e43241. doi:10.1371/journal.pone.0043241.

18 This idea is also suggested by the research of A. R. Damasio, *The Feeling of What Happens: Body and Emotion in the Making of Consciousness* (New York: Mariner Books, 2000). See also Carissa L. Philippi, Justin S. Feinstein, Sahib S. Khalsa, Antonio Damasio, Daniel Tranel, Gregory Landini, Kenneth Williford, and David Rudrauf, "Preserved Self-Awareness following Extensive Bilateral Brain Damage to the Insula, Anterior Cingulate, and Medial Prefrontal Cortices." *PLOS ONE*, 7 (2012) e38413.doi: 10.1371/journal.pone.0038413.

19 See K. J. Holyoak and P. Thagard, "Analogical Mapping by Constraint Satisfaction," *Cognitive Science* 13 (1989): 295–355; P. Thagard, *Coherence in Thought and Action* (Cambridge, MA: MIT Press, 2000).

20 For example, A. Resulaj, R. Kiani, D. M. Wolpert, and M. N. Shadlen, "Changes of Mind in Decision-Making," *Nature* 461, no. 7261 (2009): 263–66.

21 For example, A. K. Churchland and J. Ditterich, "New Advances in Understanding Decisions Among Multiple Alternatives," *Current Opinion in Neurobiology* (2012): http://www.ncbi.nlm.nih.gov/pubmed/22554881; D. Raposo, J. P. Sheppard, P. R. Schrater, and A. K. Churchland, "Multi-Sensory Decision-Making in Rats and Humans," *Journal of Neuroscience* 32, no. 11 (2012): 3726–35.

22 Read Montague, *Why Choose This Book: How We Make Decisions* (New York: Dutton, 2006).

23 See P. Glimcher, *Foundations of Neuroeconomic Analysis* (New York: Oxford University Press, 2011).

24 For a great book on this topic, see S. Blackburn, *Ruling Passions: A Theory of Practical Reasoning* (Oxford: Oxford University Press, 1998).

25 I use this example in *Braintrust: What Neuroscience Tells Us About*

Morality (Princeton, NJ: Princeton University Press, 2011), p. 172. I discuss other problems with the Golden Rule on pp. 168–73.

26 B. J. Knowlton, R. G. Morrison, J. E. Hummel, and K. J. Holyoak, "A Neurocomputational System for Relational Reasoning," *Trends in Cognitive Sciences* 16, no. 7 (2012): 373–81.

27 I would use the word *interdigitate* rather than *interweave*, save that it sounds so ghastly.

28 J. Allan. Hobson, *The Dream Drugstore: Chemically Altered States of Consciousness* (Cambridge, MA: MIT Press, 2001).

Chapter 9: The Conscious Life Examined

1 G. Tononi, *Phi: A Voyage from the Brain to the Soul* (New York: Pantheon, 2012).

2 A. Rechtschaffen and B. M. Bergmann, "Sleep Deprivation in the Rat by the Disk-over-Water Method," *Behavioral Brain Research* 69 (1995): 55–63.

3 See sleep neuroscientist Matt Walker: http://www.youtube.com/watch?v=tADK3fvD2nw.

4 For a review of the importance of sleep to memory, see S. Diekelman and J. Born, "The Memory Function of Sleep," *Nature Reviews Neuroscience* 11 (2010): 114–26.

5 C. Cirelli, "The Genetic and Molecular Regulation of Sleep: From Fruit Flies to Humans," *Nature Reviews Neuroscience* 10 (2009): 549–60.

6 C. Cirelli, "How Sleep Deprivation Affects Gene Expression in the Brain: A Review of Recent Findings," *Journal of Applied Physiology* 92, no. 1 (2002): 394–400; D. Bushey, K.A. Hughes, G. Tononi, and C. Cirelli, "Sleep, Aging, and Lifespan in *Drosophila*," *BioMedCentral Neuroscience* 11 (2010): 56. doi:10.1186/1471-2202-11-56.

7 T. Andrillon, Y. Nir, R. J. Staba, F. Ferrarelli, C. Cirelli, G. Tononi, and I. Fried, "Sleep Spindles in Humans: Insights from Intracranial EEG and Unit Recordings," *Journal of Neuroscience* 31, no. 49 (2011): 17821–34.

8 See Matt Walker discuss this: http://www.youtube.com/watch?v=giKIFuw5fyc.

9 C. L. Philippi, J. S. Feinstein, S. S. Khalsa, A. Damasio, D. Tranel, G. Landini, K. Williford, and D. Rudrauf, "Preserved Self-Awareness Following Extensive Bilateral Brain Damage to the Insula, Anterior Cingulate, and Medial Prefrontal Cortices," *PLOS ONE* 7, no. 8 (2012): e38413. doi:10.1371/journal.pone.0038413.

10 N. D. Schiff, "Central Thalamic Contributions to Arousal Regulation and Neurological Disorders of Consciousness," *Annals of the New York Academy of Sciences* 1129 (2008): 105–18; S. Laureys and N. D. Schiff, "Coma and Consciousness: Paradigms (Re)framed by Neuroimaging," *Neuroimage* 61, no. 2 (2012): 478–91.

11 L. L. Glenn and M. Steriade, "Discharge Rate and Excitability of Cortically Projecting Intralaminar Neurons During Wakefulness and Sleep States," *Journal. of Neuroscience* 2 (1982): 1387–1404.

12 Schiff gave a lecture in which he discusses the Herbert case, but also the differences between coma and vegetative state that can be seen here: http://www.youtube.com/watch?v=YIznyWtXlKo.

13 See the review article by H. Blumenfeld, "Consciousness and Epilepsy: Why Are Patients with Absence Seizures Absent?" *Progress in Brain Research* 150 (2005): 271–86.

14 B. Baars, *A Cognitive Theory of Consciousness* (Cambridge, MA: Cambridge University Press, 1989).

15 D. M. Rosenthal, "Varieties of Higher-Order Theory," in *Higher-Order Theories of Consciousness*, ed. R. J. Gennaro (Philadelphia: John Benjamins, 2004), 19–44. This self-referentiality may be important in some human conscious experience, but is not likely to be important in dolphin or raven conscious experience.

16 For a comprehensive and helpful review article, see S. Dehaene and J-P. Changeux, "Experimental and Theoretical Approaches to Conscious Processing," *Neuron* 70 (2011): 200–227.

17 See also Sebastian Sung, *Connectome: How the Brain's Wiring Makes Us Who We Are* (New York: Houghton Mifflin Harcourt, 2012).

18 T. B. Leergaard, C. C. Hilgetag, and O. Sporns, "Mapping the Connectome: Multi-Level Analysis of Brain Connectivity," *Frontiers in Neuroinformatics* 6, no. 14 (2012): PMC3340894.

19 Melanie Boly, R. Moran, M. Murphy, P. Boveroux, M. A. Bruno, Q. Noirhomme, D. Ledoux, V. Bonhomme, J. F. Brichant, G. Tononi, S. Laureys, and K. Friston, "Connectivity Changes Underlying Spectral EEG-Changes During Propofol-Induced Loss of Consciousness," *Journal of Neuroscience* 32, no. 20 (2012): 7082–90.

20 Summed up by M. T. Alkire, A. G. Hudetz, and G. Tononi, "Consciousness and Anesthesia," *Science* 322, no. 5903 (2009): 876–80.

21 See B. Heinrich, *One Man's Owl* (Princeton, NJ: Princeton University Press, 1993).

22 J. Panksepp, "Cross-Species Affective Neuroscience Decoding of the Primal Affective Experiences of Humans and Related Animals," *PlOsOne* 6, no. 8 (2011): e21236; J. Panksepp and L. Biven, *The Architecture*

of Mind: Neuroevolutionary Origins of Human Emotions (New York: W. W. Norton, 2012).

23 We saw this in Chapter 8 in the brief discussion of Dennett. See also D. C. Dennett, *Consciousness Explained* (New York: Basic Books, 1992).

24 D. L. Everett, *Language: The Cultural Tool* (New York: Pantheon, 2012).

25 N. Sollmann, T. Picht. J. P. Mäkelä, B. Meyer, F. Ringel, and S. M. Krieg, "Navigated Transcranial Magnetic Stimulation for Preoperative Language Mapping in a Patient with a Left Frontoopercular Glioblastoma," *Journal of Neurosurgery* (Oct 26, 2012).doi: 10.3171/2012.9.JNS121053 [epub ahead of print].

26 G. A. Ojemann, "Brain Organization for Language from the Perspective of Electrical Stimulation Mapping," *Behavioral and Brain Sciences* 6 (1983): 90–206; G. A. Ojemann , J. Ojemann, E. Lettich, et al., "Cortical Language Localization in Left, Dominant Hemisphere: An Electrical Stimulation Mapping Investigation in 117 Patients," *Journal of Neurosurgery* 71 (1989): 316–26.

27 M. A. Cohen, P. Cavanagh, M. M. Chun, and K. Nakayama, "The Attentional Requirements of Consciousness," *Trends in Cognitive Sciences* 16, no. 8 (2012): 411–17. For a novel and importantly different take on attention and how the brain represents attentional processes as we are conscious of events, see Michael S. A. Graziano, *Consciousness and the Social Brain* (New York: Oxford University Press, 2013).

28 Bernard J. Baars and Nicole M. Gage, eds., *Cognition, Brain, and Consciousness: Introduction to Cognitive Neuroscience* (New York: Academic Press, 2007). Michael S. A. Graziano and Sabine Kastner, "Human consciousness and its relationship to social neuroscience: A novel hypothesis." *Cognitive Neuroscience* 2 (2011): 98–113. doi:10.1080/17588928.2011.565121.

Epilogue: Balancing Act

1 Dr. Love is Paul Zak. http://www.ted.com/talks/paul_zak_trust _morality_and_oxytocin.html.

2 Another extraordinary resource for educators and anyone who is just curious can be found at the website of Howard Hughes Medical Institute. Go to their coolscience/resources, and you will find treasures galore. http://www.hhmi.org/coolscience/resources/.

3 AIDS deniers are people who think that AIDS is unrelated to the human immunodeficiency virus but is caused by recreational drugs, mal-

nutrition, and so forth. For a discussion on AIDS denialism, see Nicoli Nattrass, "The Social and Symbolic Power of AIDS Denialism," *The Skeptical Inquirer* 36, no. 4 (2012), or go to this website: http://www.csicop.org/si/show/the_social_and_symbolic_power_of_aids_denialism.

4 As claimed by Missouri congressman Todd Akin in August 2012.

5 In this connection, see the revealing article by J. P. Simmons, L. D. Nelson, and U. Simonsohn, "False-Positive Psychology: Undisclosed Flexibility in Data Collection and Analysis Allows Presenting Anything as Significant," *Psychological Science* 11 (2011): 1359–66. doi: 10.1177/0956797611417632.

6 Michael Shermer, *The Believing Brain: From Ghosts and Gods to Politics and Conspiracies—How We Construct Beliefs and Reinforce Them as Truths* (New York: Times Books, 2011).

7 To see this sort of confused lashing out in more detail, see Raymond Tallis, *Aping Mankind: Neuromania, Darwinitis, and the Misrepresentation of Humanity* (Durham, UK: Acumen Publishing, 2011).

8 Terry Sejnowski and I made this point in *The Computational Brain* (Cambridge, MA: MIT Press, 1992). Inexplicably, some people continue to assume that neuroscience is only about the level of molecules and to criticize that straw man version of neuroscience.

9 Bertrand Russell, "What I Believe" (1925), in *The Basic Writings of Bertrand Russell, 1903–1959*, ed. Robert E. Egner and Lester E. Denonn (London: Routledge, 1992), 370.

INDEX

Page numbers in *italics* refer to illustrations.
Page numbers beginning with 267 refer to end notes.

293